# Technologies of Consumer Labor

This book documents and examines the history of technology used by consumers to serve oneself. The telephone's development as a self-service technology functions as the narrative spine, beginning with the advent of rotary dialing eliminating most operator services and transforming every local connection into an instance of self-service. Today, nearly a century later, consumers manipulate 0–9 keypads on a plethora of digital machines. Throughout the book Palm employs a combination of historical, political-economic, and cultural analysis to describe how the telephone keypad was absorbed into business models across media, retail, and financial industries as the interface on everyday machines including the ATM, cell phone, and debit card reader. He argues that the naturalization of self-service telephony shaped consumers' attitudes and expectations about digital technology.

**Michael Palm** is Assistant Professor of Media and Technology Studies in the Department of Communication Studies at University of North Carolina at Chapel Hill, USA.

# Routledge Research in Cultural and Media Studies

For a full list of titles in this series, please visit www.routledge.com.

# Technologies of Consumer Labor

## A History of Self-Service

Michael Palm

Routledge
Taylor & Francis Group

LONDON AND NEW YORK

First published 2017 by Routledge

2 Park Square, Milton Park, Abingdon, Oxfordshire OX14 4RN
711 Third Avenue, New York, NY 10017

*Routledge is an imprint of the Taylor & Francis Group, an informa business*

First issued in paperback 2017

*Library of Congress Cataloging-in-Publication Data*

Names: Palm, Michael, 1973– author.
Title: Technologies of consumer labor: a history of self-service /
by Michael Palm.
Description: New York: Routledge, [2017] | Series: Routledge
research in cultural and media studies; 90 | Includes bibliographical
references and index.
Identifiers: LCCN 2016013412 (print) | LCCN 2016020900 (ebook)
Subjects: LCSH: Technological innovations—Social aspects. | Telephone
systems—Automation. | Self-service (Economics) | Service
industries—Technological innovations.
Classification: LCC HM846 .P35 2017 (print) | LCC HM846 (ebook) |
DDC 303.48/3—dc23
LC record available at https://lccn.loc.gov/2016013412

ISBN: 978-1-138-18647-7 (hbk)
ISBN: 978-0-8153-6474-0 (pbk)

Typeset in Sabon
by codeMantra

Lovingly dedicated with admiration and gratitude to the memory of Herman and Clara Margulies

# Contents

# List of Figures

# Acknowledgments

I am honored to join the ranks of authors who have acknowledged the mentorship of Andrew Ross. I continue to emulate Andrew as an advisor, a scholar, and a colleague. Toby Miller was an exacting reader and interlocutor throughout graduate school, and the phrase "consumer labor" was his suggestion. Anna McCarthy, Marita Struken, and Dan Schiller rounded out my all-star jam of a dissertation committee, each providing key insights and encouragement. In American Studies at NYU, I got schooled by a humbling array of teachers and colleagues. Along the way Pat McCreerey and Rich Blint became dear friends. Through my involvement with GSOC/UAW, I met and work with fascinating and inspiring scholars, activists, and organizers from departments across NYU and from campuses around the world. Among this wealth of comrades, Susan Valentine and Monika Krause helped me through intense periods of personal as well as professional turmoil.

At UNC, Larry Grossberg has been a mentor and a friend, and I could not ask for a better cohort of colleagues. Renee Alexander-Craft, Sarah Dempsey, Julia Haslet, Chris Lundberg, Tony Perucci, Sarah Sharma, Kumi Silva, and Neal Thomas have all buoyed me institutionally and challenged me intellectually. Tony in particular has been there whenever I've needed a hand or an ear, a meal or a drink. Every year at UNC has brought more brilliant PhD students to work with, for, and alongside, including Drs. Ali Colleen Neff, Erin Arrizi, Rolien Hoyng, and Carey Hardin. Beyond UNC, Bill Maurer and Lana Swartz helped me realize the contemporary stakes of my historical research, and Vicki Mayer has been a constant source of support. Jack Bratich, Christina Dunbar-Hester, Rebekah Moore, Jeremy Packer, and Winnie Poster stepped up with feedback and friendship at crucial moments.

It saddens me not to be able to share the publication of this book with Doc and Joan Palm, Herman and Clara Margulies, and Joe Sterbenc, but I relish the chance to thank Nancy Palm, Edward Puncher, Sidney and "baby Joey;" Jeff Sterbenc, Nicole Neff, Will, Henry and Katie; Jean Sterbenc; Ian Knox and Chad Robinson; Jina Valentine, Tom and Joan Valentine, and Sylvan Palm-Valentine. Much love and gratitude to you all, my family.

Courtney Berger, Jonathan Sterne, Eric Zinner, and four anonymous reviewers provided helpful feedback and encouragement. Finally, thank you Christina Kowalski, Felisa Salvago-Keyes, and everyone at Routledge for guiding this book down the pipeline.

# Introduction

## Phoning It in, or Consumer Labor and the Telephone

Raleigh-Durham International (RDU) is a typical midsize American airport. Outside the main terminal, several banks of trim silver kiosks line the path to short-term parking. Charge cards are inserted, then quickly removed; PINs are entered, then 'enter' is pressed. Receipts are accepted, declined, or ignored. A row of tollbooths stands guard at RDU's perimeter, armed with kiosks programmed to accept validated tickets as well as charge cards. On average self-payment kiosks cost upward of $100,000 and pay for themselves in eighteen months.[1] Annually, consumers in the U.S. spend more than two trillion dollars using self-check out terminals.[2] Recently the kiosks at RDU have been upgraded with sleeker casings and touch screens, and they have been stripped of their branding. An *ExitExpress* banner no longer hangs from the tollbooths' awning, nor does the logo adorn every machine. Elaborate instructions have also been removed. The denuded parking kiosks at RDU embody this book's animating question: how did self-service become second nature?

"It is the task of critical theory to define a new frame for understanding the everyday in the context of digital culture."[3] This charge comes from Mark Poster's penultimate book, whose title, *Information, Please*, and opening lines invoke the local telephone operator. I heed Poster's call for rethinking the everyday by narrating a history of telephone dialing through to its digital descendants. Combining historical, political-economic, and cultural analysis, I document the absorption of telephonic self-service into business models across media, retail, and financial industries. "It is when technologies such as the telephone and the computer cease to be sublime icons of mythology and enter the prosaic world of banality ... that they become important forces of social change."[4] The telephone dial was such a force in its own right. *Technologies of Consumer Labor* presents the history of telephone dialing as an origin story for the digital everyday. Ultimately this book demonstrates how the naturalization of self-service over the phone shaped consumers' attitudes and expectations about interacting with digital technology.

"Self" was first conjoined to "service" to promote a new type of discount grocery store, made popular in the U.S. during the Great Depression, but the practice of self-service predates the slogan and has always exceeded it. Take vending machines. By the time "self-service" entered the American lexicon, vending had already industrialized around "the four C's," candy, coffee, cola, and cigarettes. The number of vending machines in the U.S. climbed steadily for decades, until smoking lost its luster. Vending squarely qualifies as self-service, but its champions never bothered adopting the phrase. Today Japan boasts the highest volume of vending machines *per capita* (1:23[5]) and the widest range of products for sale, as well as the most sophisticated technology. The four C's no longer define vending, in the U.S. or Asia, and airports from Tokyo-Narita to RDU swell with elaborate kiosks displaying in-flight accouterments from neck pillows to portable media players, for rent as well as for sale. Airport ATMs are also abundant, and some old-fashioned dispensaries still accept bills and coins in exchange for newspapers, snacks, and drinks. (An early account of ATMs in *Wired* magazine heralded them as "soda machines of the future.") Outside the terminal at RDU, a Coke machine punctuates one row of parking kiosks. During my dozens of trips to and from RDU, I have never seen a beverage purchased here (despite notoriously high prices inside airports), but its bright, red presence opens a window into its neighbors' history.

*Figure I.1* A history of self-service, in two parts.[6]

The history of self-service is transnational, perhaps even global, but the slogan is patently American. Vanguard design in the contemporary vending business has become a source of nationalist pride in Japan (and Japanese

*Figures I.2–I.3* Some kiosks are better than others.[7]

corporations dominate sectors of the automobile and electronics industries, both heavily automated), but "self-service" never clicked in Japan as a slogan for automated consumption – not even in *katakana*, the alphabet of characters reserved for words imported from English. In the U.S., conversely, the first advertising campaigns for self-service stores resonated with foundational American values. To persuade customers that less service meant better service, shop owners promised an enriched shopping experience along with lower prices. Their pitch appealed to a DIY ethos that has energized Americans from Thoreau to punks. Self-service shopping promised more freedom, more privacy, and more autonomy. At the same time, of course, social status accrues to the experience of being waited on, and perhaps the most convincing evidence of self-service's historical success is the smattering of "full-service" options remaining. Increasingly self-service is the only option – or options. When self-check in was introduced at RDU, for instance, American Airlines rolled out the red carpet for its first-class flyers.

Over the course of the twentieth century, self-service became less prominent and more prevalent. Entire professions were automated into extinction, such as the local telephone operator, while other service labor faced systematic deskilling and routinization. Along the way some of the most familiar routines of everyday life in the U.S., from grocery shopping to paying for parking, were reorganized into self-services. Whether or not these routines bear the label "self-service" ultimately matters less than the degree to which their assignment to consumers has ceased to be controversial or even noticeable. Ultimately, the most successful self-service technology of all has been the idea of self-service itself. The slogan has largely been retired, and its discursive disappearance suggests that self-service has evaporated into the realm of the ordinary. The experiential familiarity of self-service for so many people belies a complex history, a history hiding in plain sight.

## From Self-Service to Consumer Labor

Self-service was a perfect fit for American infrastructure and ideology, and its naturalization became part and parcel of modern U.S. history. The slogan was first applied to unassisted shopping in grocery stores, where new economies of scale catalyzed the rise of *supermarkets*. Self-service shoppers selected their own groceries and carted them to the cash register; then, they loaded them into cars and drove them home. Supermarkets would not have been plausible, let alone flourished, without the *car culture* blossoming around them. (Even in cities with transit systems and bike cultures, a defining feature of a supermarket – along with self-service – has always been a parking lot several times larger than the selling floor.) Supermarkets sprang up across the country after World War II, as federal and municipal tax revenue famously subsidized *suburbanization* by funding the construction of bigger roads between bigger stores and bigger homes and garages. During *the postwar boom* self-service spread from stores to other venues around town, such as banks, fast food restaurants, and, of course, gas stations. Meanwhile, big-ticket technologies from cars to kitchen appliances contributed to new modes of better living through better consumption. Self-service provided an ideal link between those twin pillars of American hegemony: mass industrial production and mass consumer culture. As the practice of self-service took root, the concept congealed into common sense.

Self-service scholarship is as old as the slogan. In 1937, *The Journal of Business of the University of Chicago* published an article based on the first (and still most extensive) survey of consumer attitudes toward new self-service supermarkets and their more conventional competitors.[8] As self-service spread throughout American commercial culture, self-service studies matured into a self-referential field of business scholarship. The thrust of self-service studies has been expectantly boosterish, and at the field's cutting edge new venues have given way to new technology. SST (self-service technology) and TBSE (technology-based service encounters) are acronyms in contemporary marketing scholarship.[9] The first SST included carts and baskets, which grocery shoppers used to navigate stores that had been spatially reorganized with reconfigured aisles, shelves, and counters. Turnstiles were added to control traffic, and storage innovations at home as well as in stores were also necessary. Historically a plethora of technologies have been developed explicitly for the performance of self-service, deployed for consumers to use in concert with other, established technologies like the automobile. A technological turn within self-service studies belies the fact that self-service always has been a technological endeavor.

In their "primer" for a culture studies approach to technology, Jennifer Daryl Slack and J. Macgregor Wise elaborate the concept of assemblage as a framework broad and flexible enough to account for the complexity of anything recognizable as technological. Self-check out in supermarkets serves as one of Slack and Wise's two leading examples of technological assemblage.[10] A technological assemblage is not merely assembled from smaller pieces; although, obviously self-check out does consist of multiple material parts, including elements of other assemblages like the cash register and ATM. Also, crucially, any technological assemblage is comprised of "practices, representations, experiences, and affects [that] take a particular dynamic form with broader cultural consequences." Assemblages incorporate and, in turn, influence "habits, attitudes, ideas, and so on, which reach far beyond the effects of physical machines on cultural practices." The contours of a technological assemblage are vast, messy, and fluid, and it follows that any assemblage is "characterized by a constant process of transformation." Some of this tumult surrounding self-check out, for instance, involves recent articulations between computers, consumption and data security recasting traditional assumptions about shopping and convenience. An unpredictable ocean of ideas and attitudes "undermine[s] any assemblage's stability," making the concept especially fruitful for historical analysis of everyday technology.

Consumers' habits have always provided fertile ground for the cultivation of economic as well as ethical or ideological productivity, but self-service is systematically diffuse and historically draped in marketing. As a result, its productivity can be analytically elusive. Slack and Wise's elaboration of self-check out as a technological assemblage is especially welcome because critical self-service studies pale next to the business literature. Charles Koeber has written two essays about "consumptive labor," in which he channels Michael Burawoy to explain how "organizations manufacture consent of consumers to self-service."[11] Koeber offers the useful reminders that self-service is always a "relational phenomenon" and a "historical process," and he describes a "win-lose-lose the most" pattern among companies, consumers, and employees. Ursula Huws has researched "externalizations of labor" resulting in "unpaid consumption work," and in the sole book-length analysis of self-service, Nona Glazer focuses her sociological study on retail and health care sectors, developing the term "work transfer" to describe how the paid labor of store clerks and nurses became unpaid labor for shoppers and patients.[12] The four chapters of this book are devoted to historically pivotal "externalizations of labor" and "work transfers" from employees to their erstwhile customers. Rather than present case studies of discrete self-service technologies, one following the other, I describe four historically overlapping assemblages that share key features and fundamental elements. In order to narrate a history of self-service with case studies of technological assemblage, I adopt and elaborate the category of consumer labor. Consumer labor is essentially a synonym for self-service, but it is not constricted by the history of the slogan.

As a noun, labor receives adjectives well. Marx elaborated several distinctions (e.g., abstract/concrete, living/dead) in his labor theory of value. Since Marx, the most significant adjective for labor remains domestic. Consumer labor and domestic labor overlap, but they are neither identical nor coterminous. Consumer labor comprises a historical subset of domestic labor in which "the boundary between work … for the family and for capitalism disappears."[13] The entire process of washing clothes qualifies as domestic labor, for example, while self-service shopping for laundry detergent also entails consumer labor that historically has cut costs for soap manufacturers and merchants. Less consumer labor than domestic labor qualifies as reproductive in the Marxian sense. Thorsten Veblen's concept of "conspicuous consumption" relied on a categorical threshold beyond which consumptive desire was motivated by other factors. During Veblen's time, the domestic labor of conspicuous consumption was concealed or delegated to servants. Today trendy or cutting-edge consumer technology can function as a status symbol or marker of wealth analogous to human servants during the Victorian age. The charms of consumer technology – having "an app for that," as the Apple slogan put it – can overshadow its applications as consumer labor technology.

Contemporary critics of capitalism are less likely to describe everyday life in opposition to labor than they are to find within the everyday new modalities of labor. Lefebvre understood the everyday to be a residual category, left over after the more explicit spheres of life were all accounted for, but lately it is labor that has become "understood as a larger category with which to analyze many different facets of daily life."[14] A recent spate of neologisms like playbor and prosumption signals something like an epistemological overflowing of labor into the realms of leisure and consumption.[15] In contemporary analyses of capitalism, labor "is increasingly under pressure as an analytical category in a world where the boundaries between work and life are breaking down."[16] Christina Morini's and Andrea Fumagalli's elaboration of a "life theory of value" suggests that nowhere is humanity any longer insulated against capitalist appropriation.[17] Today, it seems, any endeavor might be mined for economic productivity. Perhaps labor theory has reached a level of saturation where the capacity for some sort of productivity can be imagined into any aspect of human life on earth, as a sort of "labor beyond labor" to twist the title of Negri's influential rereading of the *Grundrisse*. A "life theory of value" gestures toward an end game or finish line, after which no new vocabulary for labor theory is required or perhaps even possible.[18]

An abundance of new labors circulates within contemporary critical theory. Tiziana Terranova appropriated the term free labor to describe Internet users' unpaid contributions to software and content online. In the fifteen years since Terranova's groundbreaking essay, Mark Andrejevic and others have similarly mined Marx's vocabulary to critique vanguard modes of leisure and consumption.[19] Likewise, critiques of digital labor, at least to date, have tended to focus on creative and leisurely pursuits, while studies of routine digital labor remain stubbornly tied to employment.[20] However, unpaid digital labor, what Terranova called free labor, can be boring and even tedious, alienating from a Freudian as well as Marxist perspective. Self-payment kiosks no less than Twitter hash-tags are assemblages that channel free digital labor into (more) productive routines. Consumer labor shares with free labor and digital labor an emphasis on technology as integral to its cultivation, but consumer labor entails the use of simple as well as complex machines. The smart phone often qualifies as consumer labor technology, but so does the shopping cart. At the risk of (further) cluttering the field, consumer labor is a more precise and delineable category than ones like digital labor and free labor. Consumer labor does not indicate that all consumption has become (potentially) productive; rather, it specifically refers to those tasks, activities, and responsibilities that have been historically reassigned from employees to customers.

Several terms updated from Marx to describe postindustrial capitalism are associated with the political and intellectual movements referred to collectively (and loosely) as Autonomia. Much of the appeal of an Autonomist perspective stems from flipping the power dynamic inherent to the labor-capital relation. The autonomy of Autonomia is labor's, but not consumer labor's. Consumer labor is not performed by the "the general intellect," or by any of the empowered and empowering collectivities assumed by Autonomist theorists as inherent to laboring subjectivities. Agency is "possible or not possible depending on the particular assemblage ... [and] the agency active in any assemblage isn't necessarily *human* agency."[21] The agency and productivity of consumer labor can be severed as thoroughly as they can be sutured. In what some Autonomist labor theorists took to the calling "the social factory," the entire planet is characterized as a workshop in which capital and labor vie for autonomy over the latter's productive capacities. The productivity of labor-power in the social factory can never exhaustively be appropriated into capital. Consumer labor-power, however, operates in technological assemblages less akin to the social factory than to a social office.[22] It is difficult if not impossible for laboring consumers to identify with the politics described by Autonomist theorists. The agency of consumer labor in the social office operates at a deficit instead of a surplus. In *White Collar* C. Wright Mills discovered that actual office work could be as alienating as any assembly line, and he described "the white-collar people" as "above all else ... a new cast of actors, performing the major routines of twentieth-century society."[23] The history of consumer labor reveals that, all along, it has been a vaster cast of actors than Mills' white-collar people performing these routines.

## Dialing and Its Afterlives

Among proliferating access points and interface, a telecom connection of one sort or another is still required to traverse most digital networks. Originally, telecomm connections were the jobs of telephone (and telegraph) operators. Lisa Gitelman has traced the protocol formation attendant to new media, and one of her historical examples is the naturalization in the U.S. of the greeting, "Hello?" upon answering a telephone.[24] Before the dial, operators announced the identity of every caller. "Hello?" was a new protocol within the technological assemblage of rotary dialing. More recently caller ID has rendered "Hello?" functionally obsolete, while commands like "Press 1 for English" have replaced "How may I help you?" for many customer service protocols in the U.S.

Dialing was never labeled self-service, due to historical timing and especially marketing. "Ma Bell" began to roll out the dial a few years before the first self-service grocery stores popped up, and the company steadfastly avoided the phrase once it entered the lexicon. During the telephone's formative era, operators' voices accompanied the new machines into Americans' homes and offices. By "employing young, single, native-born white women to cater to bourgeois concepts about servitude," AT&T cultivated "a social and cultural relationship" between its operators and its subscribers.[25] When the telephone was new, it functioned as status symbol as well as a tool.

Initially, in fact, Bell marketed telephony as a luxury service rather than as an everyday technology. Operators were employed to provide the service, so the subsequent transition to rotary dialing was an arduous one. The Bell Company devoted vast amounts of time and resources to weaning their subscribers from operator service. Bell's customers were reluctant to say goodbye to their "hello girls," as operators were condescendingly but affectionately known, in no small part because of Bell's relentless promotion of its operator service, touted in advertisements as "the voice with a smile." Marketing and promotional materials acknowledged dialing as automation, but never as self-service. Whereas the first advertisements for self-service stores trumpeted a superior shopping experience, Bell committed to an aggressive marketing strategy for the dial that downplayed callers' new tasks and responsibilities.

Interactions between operators and callers were unprecedented in their ephemeral intimacy and gender relations. Operators' voices invisibly filled the ears of callers at home, at work, and inside phone booths, and telephone calls were the first form of men's communication over which women were granted technological authority. Given operators' iconic status historically, it is no surprise they are the first occupation mentioned in Amy Sue Bix's history of "American's debate over technological unemployment." Bix dates the debate to the Great Depression, when anxiety about automation "stretched across occupational lines to become a nationwide controversy."[26] During "the 1920s, mechanizing production had seemed to guarantee prosperity; during the 1930s, people feared that changing workplace technology might become America's social and economic downfall."[27] Yet flashes of hostility toward the dial sparked during the "roaring '20s," when enthusiasm for automation still ran high. With the rotary dial, technological unemployment spilled out of the factory, and consternation about the dial spread not only across class lines and collar colors, but broader categorical divisions as well.

Operators were the first predominantly female occupation to face automation. Dialing was also an unprecedented case of labor automation because, for the first time, it worried American citizens about their consumption as well as their jobs. In 1920, for example, the Cleveland City Council passed a resolution against the installation of a citywide dial system for local calls, because such a system would "put the burden ... upon the subscriber." Senator Carter Glass (D-Va) authored a resolution to ban the dial on Capitol Hill in 1930, on the cusp of the Great Depression (and three years before the Glass-Steagall banking reform act). The number of operators on AT&T's payroll was plummeting, but Glass' protest was not organized in support of working women. When faced with the prospect of a "roll your own" telephone being installed in his office, the senator "object[ed] to being transformed into a telephone operator [him]self without compensation."[28] Bix's history thrums with activists, organizers, and officeholders protesting automation on behalf of workers, but Glass was looking out for himself. His reaction to the dial highlights what set it apart from other formative cases of automation – the displaced workers had interacted directly with their employer's customers. Some aspects of operators' jobs were automated, while callers assumed responsibility for others. Over time though, callers forgot about the dial and began taking it for granted. Today cell phones feature full keyboards and touch screens, yet "dialing" lingers as a comprehensible verb to describe placing a call. (And its echo can be heard in the clicks, beeps, and other facsimile sounds occasionally accompanying the touching of screens.) Such discursive resiliency reflects how much marketing as well as management sustained AT&T's formidable project of automating its service.

An operator's job demanded emotional as well as mental and manual labor. Connecting calls generated value for operators' employers, and so did interactions with callers. Customer service can "create capitalist value" in its own right, above and beyond its familiar function of increasing the value of commodity goods and services.[29] Before callers began dialing telephones, operator's interactions with their customers entailed a kind of value-added that Bell used to distinguish itself from competitors. Like affective or emotional labor in this context, value-added is a contemporary concept being applied anachronistically. Yet all three terms help convey how the monopoly employed operators to imbue the technological function of connecting calls with a social and cultural relationship between servants and served. When telephone connections were automated, it proved more difficult for callers to let go of their relationship with operators than to assume responsibility for doing an operator's job.

In the wake of Taylor's infamous treatise, critical studies of automation found their subject in the factory. Labor sociologists following C. Wright Mills and Harry Braverman documented how automation's impact in offices and other white-collar workplaces could be as "degrading" and "immiserating" as any factory.[30] Labor historians following Ruth Schwartz Cohen exposed automation in the home as generating "more work for mother."[31] (Home automation assemblages often cohered around kitchen appliances perniciously promoted as "labor-saving.") Ethnographers following Arlie Hochschild specified the thesis for pink-collar service jobs.[32] Throughout most of the twentieth century, automation was a problem about the quantity and quality of jobs, while two recent popular press books suggest a shift, or expansion, from jobs to everyday life as the primary locus of automation. In *The Glass Cage* Nicholas Carr worries over the mushrooming relationship between "automation and us," while Craig Lambert recycles Ivan Illich's title *Shadow Work* to account for "the unpaid, unseen jobs that fill your day."[33] The books share the common assumption that technology has saturated contemporary living, in the process redefining the human experience of everyday tasks and routines.

Carr's premise is that computers do too many "things we used to do ourselves."[34] At stake is a surrender of knowledge and autonomy to machines and the corporations controlling them, and Carr's concerns are more psychological and ethical than occupational. Lambert is more focused than Carr and far more sanguine. Whereas Carr warns that computerized automation is becoming an unstoppable force, Lambert wants to help us better attend to this force as it reorganizes our lives into an expansive array of "unpaid tasks we do on behalf of businesses and organizations."[35] Carr sees computerized automation engulfing our lives as well as our jobs; Lambert similarly insists the "force that underlies the flood of shadow work is *"[t]echnology and robotics."*[36] Carr fears rampant computerization will leave us with nothing to do, and worse, no longer knowing how to do anything. In *The Glass Cage*, metaphorically, gender discrimination has gained four walls and a floor and mutated into a societal suppression of Weberian proportions. Shadow work, on the other hand, just "gives us more to do."[37] One of Carr's concerns is a withering of craft; Lambert is interested in a proliferation of chores, in how "consumers [as well as] robots absorb jobs," and he presents *Shadow Work* as something of a field guide for best practices.[38] Carr and Lambert both present sweeping accounts of contemporary automation, in which the history of telephone dialing hides in plain sight.

The automation of telephone calls, aka dialing, did not eliminate labor or knowledge so much as reassign each from operators to consumers. Alongside the dial, telephone directories became another new technology of automated calls. Once callers began dialing, operators were no longer needed to connect them to one another by plugging cables into a switchboard. At the same time, operators' knowledge of which cables went where on the board gave way to callers remembering some numbers and learning where to find the rest. "The phone book" became as commonplace as the phone itself, used over the years as a booster seat and doorstop as well as a municipal reference guide. Bell (later AT&T) dominated the first century of American telephony, while directories became synonymous with another brand name. Legend has it that the first yellow directory resulted from a printer's mishap, but the R.H. Donnelly Corporation claims to have published the first "official" Yellow Pages in Chicago, in 1886.[39] The yellow pages in telephone directories have been reserved ever since for businesses, alongside white and blue spin-offs for residential and governmental listings. Alongside any number of printed media, from maps to newspapers, telephone directories have seen their paper ubiquity challenged by search alternatives online. Today publishers of telephone directories like Donnelly's pursue the lion's share of their profits online. Donnelly purchased a competitor in 2005 for $4.2 billion, and *CNNMoney.com* noted their soaring stock price: "The yellow pages business is suddenly sexy. Yellow is the new black."[40]

The Yellow Pages' slogan "let your fingers do the walking" invited shoppers to precede the manual labor of travelling to stores with the mental labor of looking up their numbers and calling them. The slogan also obscures the fact that directory assistance had been an operator's job, a social service provided for callers above and beyond the technological function of connecting calls. Yellow Pages ads first encouraged the dialing public to "let your fingers do the walking" in 1962. The famous slogan promoting telephone books, and all of the listings inside, was developed by the British advertising agency Geers Gross and became their first and biggest hit across the pond. The slippery syntax simultaneously invites and commands, and the second-person pronoun is gender-inclusive, although print ads featuring the new slogan displayed female fingers exclusively. In one ad, the top of the frame cuts off a pair of walking fingers at the wrist, yet it's clear they belong to a domestic laborer. Long and slender with polished nails they cast a full-bodied shadow, with its head up heading somewhere in a hurry despite sporting heels and an apron. The slogan and imagery became successful enough to fashion into a logo, and recent iterations have seen the original ladyfingers replaced by more androgynous digits.

AT&T embarked on a decades-long multi-media campaign to promote automation as an improvement rather than a reduction of telephone service. The subsequent upgrade from dials to touch-tone keypads required virtually no marketing, in large part because callers had adjusted to dialing without operators. Trading in a rotary phone for a new touch-tone model did not eliminate the presence in one's life of an iconic service employee. Neither was the touch-tone keypad used to perform any new consumer labor – at least not right away. At first, keypads just made it a little quicker to place a call. While basic calls were simplified, however, over time telephone keypads became the interface by which any number of tasks and responsibilities have been reassigned from employees to consumers. In the U.S., everyday self-service over the phone is exemplified by the automated prompt to 'Press 1 for English.' While the introduction of dialing as standard telephone practice entailed the transfer of mental as well as manual labor from operators to callers, keypads enabled the reorganization of telephony into more complex assemblages on a grander scale. Keypads have not only engendered new uses for the telephone, from automated service menus to text messaging, but they also became a familiar presence on other digital machines such as security access panels, ATMs, and debit card readers. New machines, businesses, and even industries have harnessed the productive capacities of touch-tone data entry performed by consumers and employees alike. Before the internet, a widespread automation of everyday life was taking place via the telephone keypad.

The digitization of telecommunications infrastructure required a new interface: touch-tone keypads trigger digitally sequenced tones instead of the electronic pulses generated by turning a rotary dial. One upshot of digitization, as opposed to automation, was that consumers could now conduct transactions over the phone. Telephone calls had been used to arrange payments and negotiate prices since the days of operators and rotary dials. (Some early critics of operators worried they could eavesdrop on calls while accounts were cleared and scores were settled.) But the actual exchange of currency over the phone was not possible until monetary values could be viably, legally, and securely converted into ones and zeros. Digitization tethered retail payment to telecommunications infrastructure and interface,

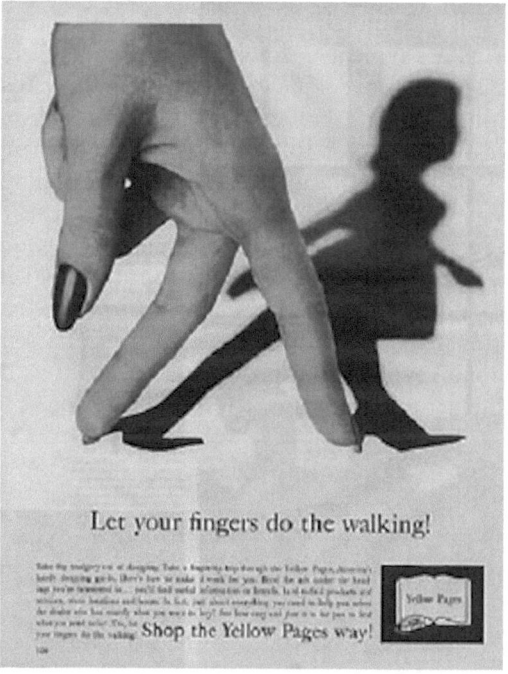

*Figures I.4–I.5* Looking it up, from walking fingers to digital surfing.

inside stores as well as remotely, and the telephone keypad has enjoyed a successful career as a payment technology. Today self-check out and self-payment assemblages feature debit card readers with keypads virtually identical to the one introduced a half-century ago on touch-tone telephones. (Unless they have a touch screen to flash a 0–9 keypad when a PIN is called for.) During the decades since the upgrade from rotary dialing to touch-tone, digital keypads have come to play a central role in the everyday communication practices of many people in the U.S. and throughout much of the wired world. The cultural resiliency of touch-tone "dialing" is demonstrated every time a keypad is represented in pixelated form. The naturalization of dialing still informs many contemporary technological assemblages, within which the telephone keypad has a proven track record as digital interface for automated interactions and transactions alike.

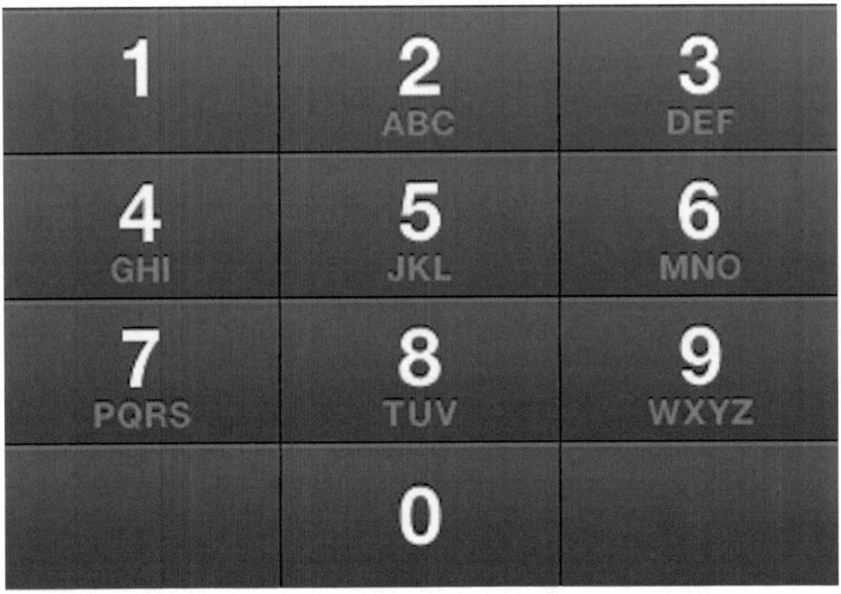

*Figure I.6* The pixelated keypad.[41]

## The Historical Narrative

There have been three key upgrades in telephone interface, corresponding with technological transformations in the telecommunication industries and society more broadly: the dial and automation, the keypad and digitization, and the touch screen and computerization. Author and ecologist Nicols Fox has described telephone dialing as "the pre-history of the Labor Transfer Movement," whereby consumers began serving themselves in everyday commercial venues – the bank, the gas station, the grocery store – until the practice became accepted and expected. Telephone keypads then catalyzed the dialing public's transformation into "unpaid receptionists for businesses everywhere."[42] The rotary dial and touch-tone keypad are technological assemblages whose histories continue to inform contemporary assemblages taking shape around touch screens and other digital interface.

Telephone dialing was a formative technological assemblage, and the bulk of this book focuses on the history of consumer labor conducted via telephones and other machines with keypad interface. Yet the history of self-service begins inside brick-and-mortar stores. In Chapter 1, "Please Help Yourself: Self-Service Shopping and the 'Revolution in Distribution,'" I describe the managerial, technological, and cultural processes by which consumers grew accustomed to serving themselves. The introduction of self-service shopping entailed myriad innovations, from mass display and uniform packaging of goods to lawsuits and legislation adjudicating consumers' new responsibilities inside stores. Accordingly, my archive for Chapter 1 features "how-to" manuals written for aspiring self-service merchants during the 1940s and 1950s, alongside social and cultural histories of shopping and legal reviews of pivotal cases. I also elaborate a political-economic analysis of self-service and its transgressions across traditional divisions between labor and consumption. Ultimately the most significant changes wrought by self-service shopping involved the cultivation of new expectations among consumers. The success story of self-service shopping is one of historical change in consumers' attitudes as well as their actions. Whether or not shoppers welcome self-service or resist it, and historically they have done both, the bottom line is that most have come to expect it.

Chapter 2, "Phantom of the Operator: Rotary Dialing and the Automation of Everyday Life," describes the establishment of dialing as routine. I recover a sense of dialing's substantial novelty by analyzing promotional films produced by the Bell Telephone Company (later AT&T). The films demonstrate how much time and energy the monopoly spent convincing its customers that dialing entailed an improvement rather than a reduction in telephone service. To capture in particular the social and cultural implications of relationships between telephone operators and their customers, I also document the attachments callers made to local exchange names, such as "BUtterfield-8," and later to area codes. Both collectively and individually, callers' personal connections to their telephone numbers bore traces of the class and status dynamics that had animated their interactions with operators. During the dial's formative era telephone subscribers struggled less with the new task of dialing than with losing their gendered, raced, and classed relationships with operators. The callers' history of dialing presented in Chapter 2 highlights how the use of consumer labor technology, even in the absence of employees, is predicated on social and commercial relations between servers and served.

Chapter 3, "Then Press #: Touch-Tone Phones and Digital Interface," begins by narrating an origin story for the touch-tone keypad. Ubiquitous today, the keypad's design was anything but inevitable. AT&T patented the touch-tone keypad in 1963, but it was not until the monopoly's divestiture during the 1980s that federal policy and legislation rendered rotary dialing functionally obsolete and left telephone subscribers with little choice but upgrading to a touch-tone phone, if they hadn't already. The onset of neoliberal tele-communications policy brought upheaval to everyday telephony in the U.S., which most consumers initially navigated and negotiated by purchasing and using their first touch-tone phones. In Chapter 3 I analyze the break-up of AT&T's monopoly by focusing on those policies and legislation that directly addressed touch-tone telephones. There are several scholarly analyses of the AT&T break-up focused on its commercial, corporate, and financial impli-cations, and they overlook the central role touch-tone phones played in con-sumers' experiences. Three new markets opened in the wake of divestiture, all of which pushed rotary phones toward extinction. Subscribers began pur-chasing their own phones, after leasing them from AT&T for a century, and touch-tone models dominated the new market in home telephones for sale. Long-distance service also became competitive, and leading upstarts such as Sprint and MCI utilized digital infrastructure that could not be accessed by the electronic pulses generated by a rotary dial. Finally, a new industry of automated telephone services, such as weather forecasts, sports scores, and stock tickers, sprang up and began capitalizing on the communicative as well as productive capacities of touch-tone data entry.

The most important development in the fifty-year history of the telephone keypad as an interface technology was its adoption on ATMs. Chapter 4, "What's in a PIN? ATMs and Keypads Beyond the Telephone," presents a case history of the ATM's swift implementation during the 1990s. By 1993, there were over one hundred thousand ATMs in the U.S. handling over $650 million in transactions annually, and by the end of the decade those numbers had more than doubled. In October 1997, a front-page headline in the *Wall Street Journal* asked its readers: "Have you noticed all of those ATMs suddenly appearing?"[43] By the following year, 1998, ATMs were handling more financial transactions in the U.S. than bank tellers, and by the turn of the twenty-first century an ATM could be found at virtually every bank branch in the U.S., as well as other everyday commercial venues where cash is spent, such as bars, bodegas, supermarkets, and street corners. The easy emergence of ATMs obscures their historical significance as a technological assemblage. ATMs were the first automated transaction technology that consumers used to access their money as well as spend it. The 0–9 keypad also appeared on ATMs just before cell phones took off, and touch screens only began to appear on smart phones after their emergence on ATMs. The inviting phrase, *touch here to learn more!*, exemplifies how ATMs functioned experientially and pro-motionally when banks began to move their business online. At the same time, the evolution of ATM cards into debit cards has radically reduced demand for cash withdrawals. The PIN has outgrown the ATM as a technology of consumer labor, and the number of ATMs in the U.S. as well as globally has crested. The telephone keypad, meanwhile, migrated onto debit card readers.

The afterlives of dialing and of self-service have merged, and in the con-clusion I bring both to bear on the recent proliferation of digital payment systems. In the first waves of self-service stores, clerks' originally expansive responsibilities were curtailed into the cashier's routine task of ringing up shoppers' selections after they had been made. In contemporary super-markets, self-check out comprises a new assemblage of consumer labor technology that reassigns the work of cashiers (and baggers) to shoppers. Self-check out is second nature for many shoppers, which belies the historical complexity of the assemblage if not the experience. Scanning and bagging one's own groceries and swiping a debit or credit card to pay for them mean that now a trip to a supermarket can be self-service from start to finish. The overall sales figure for self-check out continue to rise, but grocers are already beginning to unplug scanners and readers in anticipation of shoppers using smart phones to scan items and pay for them. The commercial emergence of interactive touch screens, coupled with the popularity of apps, has turned the smart phone into the most expansive and versatile consumer labor techno-logy to date. As digital payment protocol, swiping a debit card through a digital reader and entering a PIN onto a keypad, is giving way to swiping at a smart phone touch screen in order to complete transactions. Telephony's his-tory is as significant as any media or communication technology; in light of smart phones, it can even be argued that the telephone, rather than television, was the triumphant consumer technology of the twentieth century.[44] The telephone's history as a technology of consumer labor is ripe for excavation.

*Figure I.7* Who are you?[45]

## Notes

1. Alexander Jackson, "The costs of self-checkouts," *Baltimore Business Journal*, December 9, 2011. Accessed February 2, 2016. http://www.bizjournals.com/baltimore/print-edition/2011/12/09/the-cost-of-self-checkouts.html.

2. "Self-Checkout Usage Statistics," *Statistic Brain Research Institute*, last modified March 26, 2015. http://www.statisticbrain.com/store-self-checkout-usage-statistics/http://www.statisticbrain.com/store-self-checkout-usage-statistics/.

3. Mark Poster, *Information, Please: Culture and Politics in the Age of Digital Machines* (Durham, NC: Duke, 2006), 204.

4. Vincent Mosco, *The Digital Sublime* (Cambridge, MA: MIT Press, 2004), 7.

5. Jill Becker, "Vending machines for all your needs," *CNN.com*, last modified August 16, 2012. http://www.cnn.com/2012/08/16/travel/odd-vending-machines/.

6. Photo by author, RDU circa 2010, before the upgrade to sleeker, unbranded self-payment parking kiosks.

7. Photos by author. RDU, circa 2010. By opening with examples from RDU, I pay homage to Anna McCarthy and Sarah Sharma, both of whom conducted observational research there for their influential books about everyday technology and the cultural politics of space and time, respectively.

8. Myrtle M. Lohner, "Customer Attitude Toward Chicago Grocery-Store Practices," *The Journal of Business of the University of Chicago* 10 (1937): 233–50. It is worth noting the rarity of a female author in this journal in 1937. The formative era of self-service studies that followed lasted twenty years and was loosely bookended by: M. M. Zimmerman, "The Supermarket and the Changing Retail Structure," *Journal of Marketing* 5 (1941): 402–9; and William J. Regan, "Self-Service in Retailing," *Journal of Marketing* 24 (1960): 43; and Regan, "Full Cycle for Self-Service?" *Journal of Marketing* 4 (1961): 15.

9. See, for example, Matthew L. Meuter, Amy L. Ostrom, Robert I. Roundtree, May Jo Bitner, "Self-Service Technologies: Understanding Consumer Satisfaction with Technology-Based Service Encounters," *Journal of Marketing* 64 (2000): 50–64.

10. Jennifer Daryl Slack and J. MacGregor Wise, *Technology and Culture: A Primer*, 2nd edition (New York: Peter Lang, 2015), 157–8. All quotations this paragraph.

11. Charles Koeber, "Consumptive Labor: The Increasing Importance of Consumers in the Labor Process," *Humanity & Society* 35 (2011): 205–32; Charles S. Koeber, David W. Wright, and Elizabeth Dingler, "Self-Service in the Labor Process: Control and Consent in the Performance of 'Consumptive Labor,'" *Humanity & Society* 36 (2012): 6–29. See Michael Burawoy, *Manufacturing Consent: Changes in the Labor Process under Monopoly Capitalism* (Chicago: University of Chicago Press, 1979).

12. Ursula Huws, *The Making of a Cybertariat: Virtual Work in a Real World* (New York: Monthly Review Press, 2003): 23, 182. (Donna Haraway also uses the latter term in passing in her famous cyborg manifesto.) Nona Glazer, *Women's Paid and Unpaid Labor: The Work Transfer in Health Care and Retailing* (Philadelphia: Temple UP, 1993). See also Nona Y. Glazer, "Servants to Capital: Unpaid Domestic Labor and Paid Work," *Review of Radical Political Economics* 16 (1984): 61–87.

13. Glazer, "Servants to Captial," p. 65. See also Batya Weinbaum and Amy Bridges, "The Other Side of the Paycheck: Monopoly Capital and the Structure of

Consumption," *Monthly Review* 28 (1979): 88–103; and Ivan Illich, *Shadow Work* (London: Marion Boyars, 1981).

14. Jonathan Burston, Nick Dyer-Witheford, Allison Hearn, "Digital Labor: Workers, Authors, Citizens," *ephemera* 10 (2010): 215.

15. Playbor describes cultural production by fans, audience members and other interactive media consumers; prosumption casts a wider net inclusive of the realms of consumption beyond leisure. In Ritzer's and Jurgenson's influential essay, they claim that capitalism has always "been characterized by prosumption," while arguing that it has become "more central" during the digital or post-industrial era. In his analyses of reality TV, Mark Andrejevic occasionally uses the term "consumer labor" to describe activities that I would argue qualify as playbor, whereas the consumer labor I am describing seldom if ever does. Thanks Casey Hribar for reminding me about Andrejevic's use of the term.

16. Burston, Dyer-Witheford, Hearn, p. 215.

17. Christina Morini and Andrea Fumagalli, "Life Put to Work: Toward a Life Theory of Value, *ephemera* 10 (2010): 234–52.

18. Kathi Weeks and Peter Fleming assume a similar impasse as their starting points in recent books developing standpoints from which to critique and resist, respectively, the "biopolitics" characterizing contemporary modes of labor-power and labor management.

19. In the edited volume *Digital Labor* (Routledge, 2013), Andrejevic's chapter is titled "Estranged Free Labor."

20. Huws' body of work provides exceptions, although not her piece in the *ephemera* issue devoted to digital labor.

21. Slack and Wise, 159, emphasis in original.

22. Jack Z. Bratich has made a similar argument about a "social home." See "The Digital Touch: Craft-work as immaterial labour and ontological accumulation," *ephemera* 10 (2010): 303–18.

23. C. Wright Mills, *White Collar: The American Middle Classes* (Oxford, U.K.: Oxford University Press, 1951), ix.

24. See Lisa Gitelman, *Always Already New: Media, History and the Data of Culture* (Cambridge, MA: MIT Press, 2006).

25. Venus Green, "Goodbye, Central: Automation and the Decline of 'Personal Service' in the Bell System, 1878–1921," *Technology and Culture* 36 (1995): 919.

26. Bix, *Inventing Ourselves Out of Jobs?*, p. 5.

27. Ibid., p. 6.

28. Cleveland City Council Resolution file no. 52523, adopted by the council on June 14, 1920; "Phone Change," *Cleveland Press*, May 31, 1920; "Dial Phones, Banned in Senate, Stir House," *Washington Post*, May 23, 1930, p. 1; "Roll Your Own Phones Keep Users Arguing," *Washington Post*, May 24, 1930, p. 1, quoted in Venus Green, "Goodbye Central," p. 945, n. 90; p. 946, n. 95.

29. David Carlone, "The Contradictions of Communicative Labor in Service Work," *Communication and Critical/Cultural Studies* 5 (2008): 159.

30. Harry Braverman, *Labor and Monopoly Capitalism: The Degradation of Work in the Twentieth Century* (New York: Monthly Review Press, 1974); Mills, *White Collar*, ix.

31. Ruth Schwartz Cohen, *More Work for Mother: The Ironies of Household Technology from the Open Hearth to the Microwave* (New York: Beacon Books, 1983).

32. Arlie Hochschild, *The Managed Heart: Commercialization of Human Feeling* (Berkeley: University of California Press, 1983).
33. Nicholas Carr, *The Glass Cage: Automation and Us* (New York: Norton, 2014); Lambert, *Shadow Work: the unpaid, unseen jobs that fill your day* (Berkeley, CA: Counterpoint Press, 2015); Illich, *Shadow Work* (New York: Marion Boyers, 1981). The subtitle of *The Glass Cage* in paperback is "how our computers are changing us."
34. Carr, 1.
35. Lambert, 1.
36. Ibid., 10, emphasis in original.
37. Carr, 3.
38. Ibid., 11.
39. Mary Bellis, "History of the Yellow Pages," accessed February 3, 2016. http://inventors.about.com/od/xyzstartinventions/a/yellow_pages.htm.
40. Paul R. La Monica, "Let Your Fingers Do the Walking," *CNN Money* December 3, 2005. Accessed February 15, 2016. http://money.cnn.com/2005/12/13/news/fortune500/yellow/.
41. Screenshot by author.
42. Nicols Fox, "Volunteer Workers of the World, Unite," *New York Times*, April 9, 2005.
43. Matt Murray, "Have You Noticed All of Those ATMs Suddenly Appearing?" *Wall Street Journal*, October 7, 1997, A1.
44. Thanks Ken Hillis for the observation.
45. Photo by Jina Valentine.

## Bibliography

Andrejevic, Mark. "The Kinder, Gentler Gaze of Big Brother: Reality TV in the Era of Digital Capitalism." *New Media & Society* 4 (2002): 251–70.
Andrejevic, Mark. *Reality TV: The Work of Being Watched.* Lanham, MD: Rowman and Littlefield, 2003.
Bix, Amy Sue. *Inventing Ourselves Out of Jobs? America's Debate over Technological Unemployment, 1929–1981.* Baltimore: Johns Hopkins University Press, 2000.
Bratich, Jack Z. "The Digital Touch: Craft-work as immaterial labour and ontological accumulation." *ephemera* 10 (2010): 303–18.
Braverman, Harry. *Labor and Monopoly Capitalism: The Degradation of Work in the Twentieth Century.* New York: Monthly Review Press, 1974.
Burawoy, Michael. *Manufacturing Consent: Changes in the Labor Process under Monopoly Capitalism.* Chicago: University of Chicago Press, 1979.
Burston, Jonathan, Nick Dyer-Witheford, and Allison Hearn. "Digital Labor: Workers, Authors, Citizens." *ephemera* 10 (2010): 214–21.
Carlone, David. "The Contradictions of Communicative Labor in Service Work." *Communication and Critical/Cultural Studies* 5 (2008): 158–79.
Carr, Nicholas. *The Glass Cage: Automation and Us.* New York: Norton, 2014.
Cohen, Ruth Schwartz. *More Work for Mother: The Ironies of Household Technology from the Open Hearth to the Microwave.* New York: Beacon Books, 1983.
Fleming, Peter. *Resisting Work: the corporatization of life and its discontents.* Philadelphia: Temple University Press, 2015.

Fox, Nicols. "Volunteer Workers of the World, Unite." *New York Times*, April 9, 2005.

Gitelman, Lisa. *Always Already New: Media, History and the Data of Culture.* Cambridge, MA: MIT Press, 2006.

Glazer, Nona. *Women's Paid and Unpaid Labor: The Work Transfer in Health Care and Retailing.* Philadelphia: Temple University Press, 1993.

Glazer, Nona Y. "Servants to Capital: Unpaid Domestic Labor and Paid Work." *Review of Radical Political Economics* 16 (1984): 61–87.

Green, Venus. "Goodbye, Central: Automation and the Decline of 'Personal Service' in the Bell System, 1878–1921." *Technology and Culture* 36 (1995): 912–49.

Green, Venus. *Race on the Line: Gender, Labor and Technology in the Bell System, 1880–1980.* Durham, NC: Duke University Press, 2004.

Haraway, Donna. *Simians, Cyborgs and Women: The Reinvention of Nature.* New York; Routledge, 1991.

Hochschild, Arlie. *The Managed Heart: Commercialization of Human Feeling.* Berkeley: University of California Press, 1983.

Illich, Ivan. *Shadow Work*. London: Marion Boyars, 1981.

Koeber, Charles, "Consumptive Labor: The Increasing Importance of Consumers in the Labor Process," *Humanity & Society* 35 (2011), 205–32.

Koeber, Charles S., David W. Wright, and Elizabeth Dingler. "Self-Service in the Labor Process: Control and Consent in the Performance of 'Consumptive Labor.'" *Humanity & Society* 36 (2012), 6–29.

Lambert, Craig. *Shadow Work: the unpaid, unseen jobs that fill your day.* Berkeley, CA: Counterpoint Press, 2015.

Leverbre, Henri. *The Critique of Everyday Life.* New York: Verso, 1991.

Lohner, Myrtle M. "Customer Attitude Toward Chicago Grocery-Store Practices." *The Journal of Business of the University of Chicago* 10 (1937), 233–50.

Marx, Karl. *Capital: A Critique of Political Economy, Volume One.* New York: International Publishers, 1967.

Marx, Karl. *Grundrisse: Foundations of the Critique of Political Economy.* New York: Penguin, 1993.

McCarthy, Anna. *Ambient Television: Visual Culture and Public Space.* Durham, NC: Duke University Press, 2001.

Meuter, Matthew L., Amy L. Ostrom, Robert I. Roundtree, and May Jo Bitner. "Self-Service Technologies: Understanding Consumer Satisfaction with Technology-Based Service Encounters," *Journal of Marketing* 64 (2000): 50–64.

Mills, C. Wright. *White Collar: The American Middle Classes.* Oxford, U.K.: Oxford University Press, 1951.

Morini, Christina and Andrea Fumagalli. "Life Put to Work: Toward a Life Theory of Value. *Ephemera* 10 (2010): 234–52.

Mosco, Vincent. *The Digital Sublime.* Cambridge, MA: MIT Press, 2004.

Murray, Matt. "Have You Noticed All of Those ATMs Suddenly Appearing?" *Wall Street Journal*, October 7, 1997.

Negri, Antonio. *Marx beyond Marx: Lessons on the Grundrisse.* Oakland: AK Press, 1978.

Poster, Mark. *Information, Please: Culture and Politics in the Age of Digital Machines.* Durham, NC: Duke University Press, 2006.

Regan, William J. "Self-Service in Retailing." *Journal of Marketing* 24 (1960): 43.

Regan, William J. "Full Cycle for Self-Service?" *Journal of Marketing* 25 (1961): 15.

Ritzer, George and Nathan Jurgenson. "Production, Consumption, Prosumption: The Nature of Capitalism in the Age of the Digital 'Prosumer.'" *Journal of Consumer Culture* 10 (2010): 13–36.

Scholz, Trebor, ed. *Digital Labor: The Internet as Playground and Factory*. New York: Routledge, 2013.

Sharma, Sarah. *In the Meantime: Temporality and Cultural Politics*. Durham, NC: Duke UP, 2014.

Slack, Jennifer Daryl and J. MacGregor Wise. *Technology and Culture: A Primer*, 2nd edition. New York: Peter Lang, 2015.

Veblen, Thorstein. *The Theory of the Leisure Class*. New York: Macmillan, 1899.

Weber, Max. *The Protestant Ethic and the Spirit of Capitalism,* trans. Talcott Parsons. London: Unwin Hyman, 1930.

Weeks, Kathi, *The Problem with Work: Feminism, Marxism, Antiwork Politics and Postwork Imaginaries*. Durham, NC: Duke University Press, 2012.

Weinbaum, Batya and Amy Bridges. "The Other Side of the Paycheck: Monopoly Capital and the Structure of Consumption." *Monthly Review* 28 (1979): 88–103.

Zimmerman, M. M. "The Supermarket and the Changing Retail Structure." *Journal of Marketing* 5 (1941): 402–9.

# 1  Please Help Yourself

## Self-Service Shopping and the "Revolution in Distribution"

In *Double Indemnity,* the dupe confesses into a Dictaphone.[1] The opening scene finds Walter Neff, played by Fred MacMurray, lurching out of his car toward an office building. Access codes and card swipes are decades away, so he bangs on the big glass doors for the night watchman to let him in. Next Neff encounters the elevator attendant and does his best to quash any small talk during their ride up together. Alone at last, Neff opens his overcoat to reveal a bullet wound in the shoulder. He collapses into his office chair, lights a cigarette, and inserts an acetate cylinder into the Dictaphone on his desk. "Office memorandum. Walter Neff to Barton Keyes, claims manager. Los Angeles, July 16, 1938. Dear Keyes, I suppose you'll call this a confession when you hear it. ..." His monologue leads into flashback, a *film noir* convention. The Dictaphone was also familiar to audiences, having been trademarked in 1907 (after being developed by Alexander Graham Bell). In this context, however, it provides a twist. The plot of *Double Indemnity* unfolds via the first automated confession in Hollywood history. Along the way we learn that Walter Neff's office is typical is most every way. He and the other insurance salesmen are supported by abundant secretaries, and Neff's confession is predicated on their absent presence. He plans to get away before a secretary transcribes his tale.

The Dictaphone didn't displace secretaries, but it deskilled them by rendering shorthand obsolete. The automation of stenography also rationalized secretaries' labor. Since they no longer needed to drop whatever they doing when a superior demanded dictation, their time management and work rhythms could be supervised more efficiently. The Dictaphone depersonalized as well as routinized stenography by inserting a layer of technological mediation between the Walter Neffs of the business world and their secretaries. For better or worse, each began using the same machinery one after the other instead of interacting directly. In offices as well as factories, automation enjoyed a golden age during the first two decades of the twentieth century. Occasionally new technology eliminated jobs outright, like the rotary telephone dial and operators. More often than not, however, automation routinized the remaining labor of office employees. Sharon Hartman Strom (among others) has documented "the conjunction of mechanization, scientific management and the hiring of women as clerical workers" during this period that led to the entrenched malaise of office work later captured by C. Wright Mills in *White Collar*.[2]

Three reels after Walter Neff begins narrating *Double Indemnity*, he and his *femme fatale* have cooked up their scheme. "You know that big market up on Los Feliz, Keyes? That's the place Phyllis and I had picked as a meeting. ... We had to be very careful from now on. We couldn't let anybody see us together." Inside the "big market," Walter and Phyllis could count on being left alone. An employee never appears during their two meetings inside the store, which last about six minutes total. Other shoppers surround Walter and Phyllis but pay them no mind, with one exception – in a moment of comic relief, a short woman asks Neff to reach her some baby food and after thanking him mutters, "I don't know why they always put what I want on the top shelf." Audiences were still adjusting to self-service stores in their own lives and could recognize, if not relate to, the woman's frustration. The first self-service markets cropped up in southern California, where *Double Indemnity* takes place, and in 1944 when the film was released (and even more so in 1938, when it was set) self-service shopping was still brand new. The automation of stenography changed office work for secretaries and for insurance salesmen like Walter Neff, but the holders of insurance policies were unaffected. In workplaces where employees serve customers, rather than bosses, the introduction of new technology requires labor management of consumers as (well as) employees. The cash register, for instance, remediated more than retail transactions. It also affected how shoppers interacted with store staff and with one another.

*Figure 1.1* Shop, reverse-shop.

The bulk of consumer labor assemblages in the U.S. today entail telecommunication networks. Accordingly, *Technologies of Consumer Labor* is anchored by the home telephone, which over the course of the twentieth century became our most common and versatile consumer labor technology. Yet the history of consumption organized under the banner of "self-service" begins inside brick-and-mortar stores. The first self-service assemblages featured consumer labor technologies still used today in essentially their original form, such as shopping baskets and carts. Scholars of service work have pointed out the limitations of Harry Braverman's influential critique of automation, among them a focus on labor-management relations to the exclusion of interactions between employees and their customers.[3] In self-service stores Braverman's critique was both anticipated and turned on its head. Clerks' labor was deskilled and degraded, while consumers were subject to what Paul du Gay called a corresponding "en-skilling."[4]

During the Great Depression, in exchange for the promise of lower prices, grocery shoppers began selecting and retrieving their own merchandise unassisted. Grocery clerks' originally expansive responsibilities were curtailed into the routine task of ringing up whatever shoppers brought to the cash register. The economic impact of self-service shopping was initially one of scale. Beyond the labor costs savings of deskilling clerks into cashiers, self-service shopping became the ideal means of distribution for linking mass production to mass consumption, a cheaper and more efficient way for shoppers to get all their stuff home from the store. Trailblazing grocers discovered as much as invented self-service, by capitalizing on principles, practices, and possibilities already existent within industrial capitalism. They combined innovation and "progress" at several levels, including the popularization of consumer technology like the automobile and refrigerator alongside established labor management strategies involving automation and deskilling. Within these broader historical contexts, this chapter describes the discursive and experiential emergence of self-service.

When self-service was new, advertisements promoted the method of shopping itself. The first self-service stores, as well as the products for sale inside, became the subject of national mass marketing campaigns. Some ads promoted the new shopping experience as superior to being served – more independent, autonomous, and efficient – while most marketing characterized self-service work as fair exchange for greater savings. Beginning in the 1920s, and gaining steam during the Great Depression, shoppers by and large accepted their new tasks and responsibilities as part of what Sharon Zukin has dubbed, in her history of shopping, "the new bargain culture."[5] Self-service not only offered a rationalized method for the mass distribution of commodities; over time, it also became a reason to manufacture other commodities, big-ticket items like ATMs and self-payment kiosks purchased by merchants and service providers. Other

consumer technologies, from cars and refrigerators to cell phones and lap-
tops, are sold directly to consumers who use them for labor as well as leisure, prosumption as well as playbor, for routine tasks in the social office
and factory alike.

Many tasks and activities performed via consumer technology qualify
as consumer labor. Their history includes, if not begins with, self-service
shopping inside "big markets" like the one in *Double Indemnity*. In this
chapter's first two sections, I track the discursive emergence of the self-
service shopping assemblage, first in how-to manuals written by pioneering grocers for other grocers and then in retail and marketing scholarship
devoted to supermarkets. As self-service spread across retail sectors and
throughout service industries, many of the duties and obligations of retail
distribution were transferred from clerks to consumers, legally as well as
technologically and managerially. Shoppers began handling merchandise
before paying for it, and the management of this new interval required new
customs, rules and even laws. Some employee protections were extended
to shoppers inside self-service stores, but shoppers assumed more liability than merchants for injury and damages. In self-service stores the merchandise on display was for the first time brought out from behind the
counter, and this spatial reorganization of stores introduced into retail
a new gap between possession and sale. A physical distance as well as
temporal lag emerged between "shopping as choosing" and "shopping as
making a purchase."[6] Several formative lawsuits, reviewed in the third
section, entrenched consumers' responsibility for goods that they had
assumed possession of but not yet purchased. The final section elaborates
the cash register's pivotal role within the history of self-service shopping
assemblages, before the conclusion returns to the contemporary supermarket and to the self-check out assemblage.

## Setting the Table

In the decades between the Civil War and the First World War, a pattern
of retail shopping emerged in the U.S. organized around attentive service,
store credit and home delivery. Clerks were not employed to take your
money; they were there to handle the merchandise. They were also there
to help shoppers decide what they wanted. The job of sales clerk in this
formative era of retail service was a skilled position, one charged with
full responsibility for the customer's satisfaction and imbued with the
expertise to ensure it. Sporadic efforts to implement "impersonal selling," "display merchandising," or "unattended retail" date to the nineteenth century, but retail service featuring knowledgeable, available clerks
remained the dominant format for selling virtually everything. First in
large cities and then rapidly spreading across the country, "Mom and
Pop" stores (regularly staffed by unpaid family members) and independent merchants lost market share to department stores, such as Filene's

in Boston and Marshall Field's in Chicago, which during the 1880s began selling an exponentially increased volume and variety of goods. Customer service expanded along with the inventories and economies of scale, while credit and delivery services expanded to include installment payment plans, layaway, and return policies. Inside department store, clerks remained responsible for helping customers make their selections, for locating and retrieving goods, for charging the customer the proper amount, and for organizing delivery.

Self-service shopping was anathema to department stores. Service in department stores helped set them apart as elite or luxurious, in distinction from variety stores and five-and-dimes, such as Woolworth's, which also emerged during the late nineteenth century.[7] The first self-service stores sold groceries, seemingly one of the few genres of merchandise not for sale in department stores. The two new venues for retail consumption polarized American retail into "going shopping – an open-ended, pleasurable, perhaps transgressive experience – and doing the shopping, a regular task to be done with the minimum expenditure of time, labour and money."[8] Susan Porter Benson began *Counter Cultures,* her 1984 study of department store clerks, expecting to extend Braverman's deskilling thesis from the shop floor to the selling floor, but she found that the class and gender dynamics of training and self-presentation among department store clerks were too complicated for her to adhere to his model, in large part because rationalized labor control was disrupted by the presence of a third party, customers "going shopping." Here I reverse Benson's approach to analyze the labor management of consumers "doing the shopping" within the first self-service assemblages, namely grocery stores and supermarkets.

Three retail innovations from the late nineteenth century helped pave the way for self-service stores. First, during the 1890s, Sears-Roebuck pioneered mail order shopping. The mail-order catalog relocated "going shopping" from the store to the home. With a mail-order catalog shoppers could not handle the merchandise before buying it, but they could browse at their leisure, unassisted, left alone with the goods for sale – or, rather, with advertisements for them. Familiarity with goods via mass advertising was one of the commercial conditions of possibility necessary for self-service stores to succeed. Rather than asking clerks for help, shoppers would either need to know what they wanted when they arrived at the store, or have the confidence and wherewithal to decide once inside. Without the assistance of clerks in stores, self-service shopping would require name brand mass advertising in their place.

Between 1893 and 1894, the Sears Catalog expanded from less than 200 pages to over 500 pages, and by 1901 the catalogs were so popular that Sears began charging 50 cents for one, while continuing to distribute them for free to "preferred customers."[9] The practice of separating customers by status continues today in any number of forms, from frequent flyer points systems to discount loyalty cards to punch cards promising rewards. Furthermore, the fact that Sears bestowed "preferred" status on customers who continued to receive the same service as before, in this case a free catalog, anticipated the widespread practice today of reintroducing "full-service" features as perks, benefits or status symbols once customers have acclimated to self-service.

Alongside the mail-order catalog, two other new retail forms – the vending machine and the cafeteria – helped prepare consumers for self-service shopping. The vending machine was a break-through in automated shopping as well as arguably the origin. The first known design for a vending machine comes from first century Greece, when the mathematician Hero "described and illustrated a coin-operated devise to be used for vending sacrificial water in Egyptian temples."[10] A second antecedent to modern vending occurred in seventeenth- century English pubs, where patrons passed snuffboxes that could be opened by inserting a coin. Patrons had access to all of the contents inside, not unlike contemporary "honor boxes" selling newspapers, but the snuffboxes never were out of sight for owners and employees (and other patrons). There is no evidence that pilfering was a problem in the pubs, but two centuries later theft would be a primary concern for the proprietors of early self-service shops.

Commercial vending began in earnest during the 1880s, when the Adams Gum Company (later American Chicle) began selling Tutti-Frutti gum on New York City subway platforms. Gum was an ideal product to offer for sale in vending machines, for several reasons: it was a low-priced impulse buy easily consumed at the point of purchase, it withstood fluctuations and extremes in temperature (unlike chocolate, for instance), it had a long shelf life, and it bore no significant health concerns for consumers. For roughly thirty years, vending machines dispensed gum and penny candy and little else. The breakthrough for vending machines came in the 1920s, when for the first time a pulled lever dropped a pack of cigarettes. For more than half a century cigarettes would continue to be the most popular item purchased via vending machines, until the number of smokers in the U.S. began to decline. During the 1930s, soft drinks joined the line-up, and automatic coffee machines emerged during the 1940s. After World War II, the vending machine industry came to be dominated by the "4 C's" – cigarettes, cola, coffee, and candy.

Grand visions – utopian and dystopian – about vending are as old as the machines. As soon as the cigarette machine took off, "the industry and the general media would focus on the idea that 'robots' could sell just about anything and were the future."[11] Some retailers began making impassioned defenses of flesh-and-blood salespeople. One characteristic guide to "retail selling and store management," published in 1925, acknowledged that vending machines were acceptable for selling cheap, trivial items like the 4 C's, but that for any commerce beyond that scope, "[t]here is no mechanical devise that can take the place of the real salesman [*sic*]."[12] When this proclamation was made, the first experiments in self-service stores had proven successful, although the phenomenon had yet to sweep the country and become the standard method of grocery shopping. Supermarkets would be commonplace in less than twenty years, as would self-service shopping in other retail sectors, and at least one historian of shopping has concluded that "[t]he alternative to the living salesman was not, in the end, the machine; it was self-service."[13] Self-service and vending machines are hardly distinct, and a champion of vending machines proclaimed in 1945 that "[t]he ultimate perfection in self-service is provided through the vending machine, and all progress in the field of self-service will lead to wider and wider use of coin-operated devises to sell goods."[14] Neither vending machines nor self-service *writ large* can function without employees. Behind-the-scenes labor is still required to stock the shelves of a self-service store or the racks of a vending machine, not to mention ATMs and Amazon.com warehouses. Yet the absence of employees *while* consumers select, retrieve and pay for merchandise stored in vending machines marks a significant threshold (if not an "ultimate perfection") for self-service.

The mail-order catalog and the vending machine are two historically significant assemblages in the history of self-service. Each predates the discursive emergence of self-service; at the same time, vending machines share parts and protocols with self-payment kiosks, and online shopping is essentially email-order. The assemblage most closely related historically to the self-service grocery store, however, is the cafeteria. Self-service shopping extended the cafeteria's sale of prepared ready-to-eat food to packaged food, taken home and stored by shoppers for later consumption. Automated food service began in Germany during the 1890s, and the first fully mechanized "Automat" restaurant was opened in Philadelphia in 1902. Diners selected their food and opened a glass case by inserting money into the proper slot. As with vending machines, paid workers were still required to stock the cases, and most Automats served only "supplementary items such as cake, sandwiches, drinks and so on through a machine. ... Full main meals were never served in Automats through machines."[15] The formative innovation of retail distribution in the automats was the same as that in cafeterias – not the elimination of paid labor, but rather the naturalization of customers assuming responsibility for new aspects of service, in this case busing their own food to their tables and then clearing the tables when finished eating.

The phrase "self-service" was first used to describe stores in 1912, as cafeterias and vending machines were becoming more commonplace. Most grocers were wary of doing away with clerk service in one fell swoop, and often a "duplex arrangement" was implemented: "The duplex store was divided by a shelf, approximately 5 feet high, running down the middle. On one side, clerks waited on customers; the other side was self-service" and cash-and-carry rather than credit and delivery.[16] Nearly all of these duplex stores were located in the Southwest and, to a lesser extent, the

Pacific Northwest, and "Southern California is generally referred to as the cradle of the large market."[17] The automobile was becoming more popular, and dispersed populations in the West were already growing accustomed to driving long distances to buy provisions in bulk. The popularity of the automobile, coupled with climate extremes of heat or rain, led merchants throughout the U.S. West to experiment with expedited forms of retail. (Drive-thru windows, for example, became very popular across the country.)

Also in 1912, Lutey's Marketeria opened in Butte, Montana. The Lutey Brothers transplanted the cafeteria system to the grocery store. Like the duplex arrangements in California, they opened a self-service Marketeria adjacent to their full-service store, which was already successful and well-established, complete with a mail-order business and a wholesale warehouse as well as specialty departments including a coffee roaster and a bakery. Next door in the new Marketeria, shoppers encountered lower prices and paid cash for goods rather than charging them. Clerks handled the cash and wrapped the goods, which were then still delivered home. The Marketeria also featured what the Lutey Brothers advertised as "personal shopping," which meant that shoppers walked unattended through aisles of prepackaged goods, selecting what they wanted, and carrying their haul to the checkout stand. As in the duplex stores in California, woven baskets were available by the front door for shoppers to collect their haul and carry it to the check out counter.[18]

One early visitor to the Marketeria was Clarence Saunders, who returned home to Memphis, Tennessee, and founded what would become the first chain of self-service stores. Saunders' first Piggly Wiggly store opened in 1916, and like the Luteys, he took advantage of his existing wholesale business to prepackage goods uniformly, in order to ease customers' selection and handling of them. Unlike the Marketeria, however, Piggly Wiggly stores accepted only cash, no credit or checks, and no telephone or mail orders were accepted. Whereas the Lutey Brothers had located their experiment in "personal shopping" literally next door to a full-service counterpart, Saunders designed his new store to minimize the service options therein. The Piggly Wiggly chain provided retailers nationwide with the first demonstration that merchants could get along without clerks and that shoppers wouldn't miss them. The immediate success of Saunders' new system "is generally credited with setting off the [self-service] trend."[19] Saunders devised the first full-fledged self-service system: entrance and exit turnstiles and a one-way aisle system meant that shoppers passed every item for sale in the store before proceeding to the checkout counter. Photographs of the first Piggly Wiggly store in Memphis show shoppers filing single-file back and forth through aisles on both sides of the turnstiles. Shopping baskets were stored next to the entrance turnstile, and above the baskets was a sign reading "Patented October 9, 1917," referring to C. Saunders Self-Service Store, U.S. Patent No. 1,242,872.

*Figures 1.2–1.3* The Piggly Wiggly way.[20]

The Piggly Wiggly store was the first patented assemblage to channel shoppers into something called self-service, and "the secret of the operation was a turnstile at the checkout counter."[21] The Piggly Wiggly turnstiles highlighted the store's novelty: they functioned as a threshold, alerting shoppers to something different waiting for them on the other side. And once inside, there was literally no turning back. The policing function performed by turnstiles was less about minimizing theft than maximizing purchases, since a one-way entrance forced shoppers to pass by all the merchandise for sale before reaching the exit. Before long, however, the turnstiles became unnecessary, even counterproductive. By 1940, turnstiles were on the wane. A how-to guide for food merchants seeking to convert their stores to self-service argued that turnstiles were no longer necessary because "they solve no problem that cannot be solved by some other more desirable means."[22] When Saunders patented the floor plan for his first Piggly Wiggly store, he wanted to exercise strict control over shoppers. Once shoppers had grown accustomed to serving themselves, control could be relaxed.

Like any technological assemblage, self-service shopping was a combination of old and new. Merchants promoting self-service took advantage of existing technologies, like the turnstile, and the successful implementation of self-service shopping "depended on earlier innovations in packaging as well as on new technology and store design."[23] Furthermore, while self-service shopping occurred inside stores, "changes in central control (such as through chains) and in production and merchandising explain its widespread success."[24] The first Piggly Wiggly stores were less the origin of self-service than its tipping point. In *The Control Revolution*, James Beniger argues that what set Saunders' design apart from other innovations in the "retail control of distribution" was that his "essential idea was to process neither transactions nor commodities as his primary retail function but rather customers themselves." The novelty of Saunders' floor plan, and what made it worth patenting (unlike the sporadic self-service markets in the West) was that it was "explicitly designed to process people past merchandise."[25] Saunders' processing of people rather than merchandise or transactions highlights that, from the get-go, the most productive self-service technology of all has been the self-serving consumer. In order for self-service to work, shoppers "ha[d] to be taught a whole new attitude toward purchasing, and a whole new model of how to purchase. ... The assumption that the customer is 'waited on' must be disarticulated, and the customer must be convinced that this is a good thing."[26] Self-service shopping was a new assemblage requiring new architecture, new customs, and a new mindset among consumers.

The physical guidance of shoppers was only one element of the new self-service assemblage. The placement of goods and their presentation also underwent significant changes. And while Saunders' patented design is recognized as the first self-service store, several elements of that system predated Saunders' patent and contributed to it. For example, by 1909 Woolworth's "stopped building storage shelves behind the counters, and moved all items to counter tops where shoppers could reach them without asking a sales clerk for help." In 1913 A&P, the first chain of grocery stores, did the same, and "within five years, nearly all grocery stores followed this example."[27] Saunders capitalized on the shelving strategy already being used by Woolworth's, A&P and other grocers. By adding turnstiles and a one-way aisle system to ensure that shoppers had to pass by each item for sale in the store to get from the entrance to the exit, Saunders did not introduce self-service so much as perfect it.

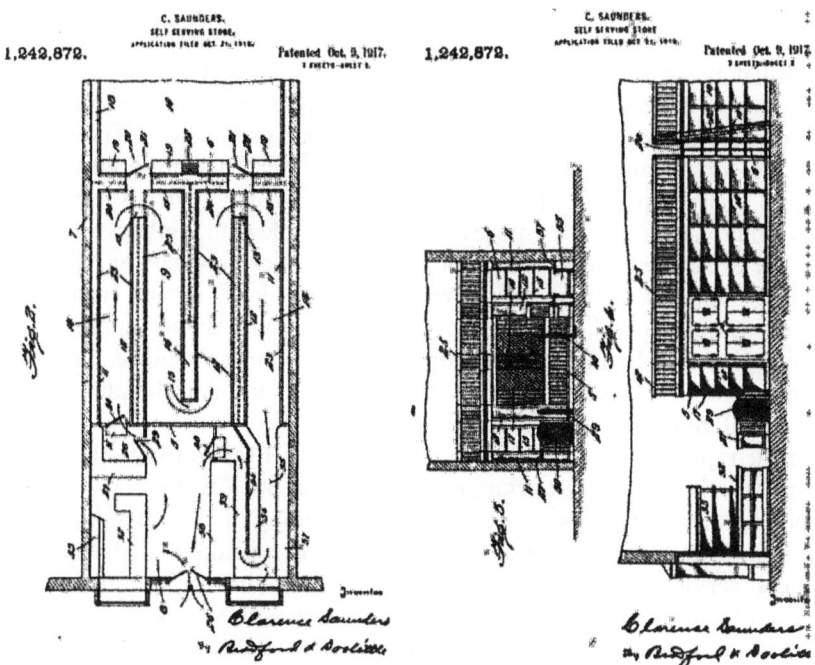

*Figure 1.4* The patented Piggly Wiggly design.

## Birth of the Supermarket

> The modern store attempts to encourage customers to do
> as much of the work of serving themselves as possible.[28]

In 1940 *The Progressive Grocer*, the "national magazine of the grocery trade," published the first how-to book for grocers seeking to convert their stores to self-service.[29] *Self-Service and Semi-Self-Service Food Stores* includes chapters devoted to "Principles of Store Arrangement," "Shelving That Invites Self-Help" and "Arranging the Grocery Department Floor For Self-Help." The editors elaborate a distinction between "self-service arrangement" and "self-service operation" and insist "[i]t is of course impossible to carry on a self-service operation without a self-service arrangement."[30] The two primary elements of a self-service arrangement are the shelves and the floor plan. The shelves of new self-service stores were not only designed to facilitate easy handling of merchandise; their design was a matter of scale as well. "Mass has a magic selling power. It is better to have big full stocks on the selling floor and shelves than skimpy stocks with surplus good piled in the back room."[31] The mass display of newly designed shelves was not only promotional, but also a newly rationalized form of storage: "[i]t is a growing practice among progressive food merchants to provide enough storage space in center shelves (as well as in side-wall shelves) so there need be little surplus stock … every shipment of goods can be placed in stock immediately."[32] Once redesigned, the shelves of a self-service store were also reorganized, as was the relationship between departments of the store: "the grocery department should be placed in the rear of the store. The greater demand for a particular grocery item, the further back it should go."[33] Staples such as milk and butter were often placed far apart so that shoppers would have to pass by more merchandise while walking between them. Logical exceptions were made, and despite the development of "gliders," as early shopping carts were known, bulk items like flour and sugar were placed near the checkout counter, so shoppers could avoid lugging them around the store by picking them up last. Bulky items, of course, also took up more room in a shopping cart or basket, discouraging the purchase of additional items.

If the repositioning of goods was a strategy for facilitating self-service shopping, then uniform packaging was a necessary element. "Self-service depended on customers and checkout staff being able to handle goods easily and identify them; without these two conditions being met, customers' service of themselves would have been virtually impossible."[34] The uniform packaging found in self-service stores was the culmination of over fifty years of development in manufacturing technology. Machines were not only increasingly mass-producing goods for sale, they were also making the packages containing mass-produced goods. "The first paper-bag-making machine was patented in 1852 … [and] in the late 1860s, a patent was granted for the now familiar square-bottomed bag."[35] Paper bags became a means of packaging goods for sale, such as flour and sugar. In self-service stores the bags also became advertising for stores themselves, emblazed with the store's name and used to bundle shoppers' hauls. After paper bags came cardboard boxes, then tin cans and glass bottles. Cardboard cartons for items like biscuits and cereal began in the 1880s to be produced by machines, and the manufacture of new packaging materials continued apace. Perhaps the most significant development was the advent of automated canning. "The first 'automatic-line' canning factory brought the continuous-process concept to canning in 1883, with machinery that could make cans at the rate of about 3,000 per hour." After the turn of the century, developments in the mass-manufacture of packaging materials continued apace: "fully automatic bottle making after 1903, aluminum foil after 1910, and cellophane, first manufactured in France in 1913."[36] As packaging materials proliferated, increasingly they were also used to promote the merchandise inside.[37]

The uniform packaging of goods was more than a matter of advertising; it was central to the rise of mass production: "Manufacturing companies that chose to invest in the new machinery and purchase the new packaging materials found themselves literally making a new kind of product."[38] Uniform packaging was also a means of standardization. "If goods were to be sold via self-service, transaction processing could be speeded up and yet controlled through the preprocessing of packaging: standardized sizes and weights for standardized prices."[39] Prepackaging in self-service stores was a central component of "the movement toward standardization, a crucial term in American debates of the 1920s and 1930s about mass production and mass consumption."[40] Standardized packaging meant the benefits of self-service shopping were not limited to reduced labor costs for employers. The value of self-service was a matter of scale as well. With uniform packaging and mass display as its twin engines, self-service shopping exponentially increased the volume of trade in retail sales and accelerated the turnover of merchandise.

New store design and the uniformed packaging of goods – a "self-service arrangement" – helped contribute to the successful implementation of self-service shopping – a "self-service operation." And, with these changes in place, the deceptively simple act of pulling goods from the shelves of a store, rather than asking a clerk to retrieve them for you, led to arguably the most significant retail innovation of the twentieth century: the supermarket. M. M. Zimmerman founded a new business journal, *Super Market Merchandising*, in 1936 and called self-service "the *sine qua non* of the supermarket operation."[41] Given the postwar proliferation of self-service beyond the supermarket into nearly all sectors of retail, it is no stretch to claim that self-service itself is a greater innovation than the supermarket. But it was, without question, in the supermarket where self-service blossomed as a new retail system. Not only was Zimmerman the founder and first editor of *Super Market Merchandising*, he also founded the Super Market Institute in 1937, which joined together "owners of supermarkets from across the United States ... to educate each other; study economic problems in the trade; to be a clearinghouse for information; and to promote relations with manufacturers, distributors, government, labor, and the public."[42] Zimmerman was "both perpetual propagandist and perpetual historian of new developments in food retailing; the supermarket was his pre-war cub and post-war giant."[43] (I took the subtitle for this chapter from the title for Zimmerman's triumphal account of the supermarket's ascent.[44]) He was as much a champion as an analyst of the early supermarket industry, but Zimmerman's knowledge of the supermarket business was second to none, and he seldom missed an opportunity to stress the centrality of self-service to the business he loved. "Self-service had, of course, been tried before the advent of the supermarket, but never under such Elysian conditions."[45] Self-service was central to Zimmerman's very definition of a supermarket: "A Super Market is a highly decentralized retail establishment, either wholly owned or concession operated, with adequate parking space, doing a minimum of $250,000 annually [in 1955]. The grocery department, however, *must be on a self-service basis*."[46] The reasons for this insistence involve the magnitude of scale mentioned already. As Zimmerman understood as well as anyone, self-service was the "paradigm shift" that enabled the supermarket to launch a "revolution in distribution" during the golden age(s) of mass production and mass consumption in the U.S.[47]

Over the course of the 1930s the standard mode of food shopping in the U.S. flipped "from the long-established habit of service, credit and delivery to that of self-service, cash and carry."[48] During the Great Depression, most shoppers were open to any new means of stretching their "food dollar," and the supermarket chains promised greater savings in exchange for self-service shopping. Even as enthusiastic a chronicler as Zimmerman acknowledges "the [supermarket] movement needed the impetus of a depression to accelerate its growth."[49] In 1932, there were less than 300 supermarkets in the U.S., doing approximately $1–1.5 million in business. The depression brought supermarkets to the East Coast, where "Big Bear" stores in New Jersey and "King Kullen" supermarkets on Long Island became immediately successful. In the East, supermarkets demonstrated an ability to thrive among denser populations and intense competition. The supermarket's success in the industrial centers of the Northeast is what "transformed self-service retailing from an isolated phenomenon on the Pacific Coast to a nation-wide movement." By the end of the decade, there were nearly 8000 supermarkets in the U.S., found in all forty-eight states and doing an annual business collectively of over $2 billion. In a more sober supermarket history, James M. Mayo offers a corrective to Zimmerman's language and tempers his tone, arguing that "the supermarket was not so much a physical design revolution as it was an economic evolution" of grocery stores chains.[50] Retail and manufacturing developments culminated in self-service shopping, while macro-economic conditions helped expand self-service grocery chains, like Piggly Wiggly, into national corporations.

Within the self-service shopping assemblage, agency was redistributed from merchants as well as shoppers toward corporations overseeing supermarkets and the mass production of groceries sold inside them. William H. Albers, of Albers Super Markets, stressed in his inaugural address as the Super Market Institute's first president, "the need for supermarket owners to promote the sale of nationally advertised brands."[51] In addition to standardized packaging – and in tandem with it – another prominent change necessary for self-service shopping was the rise of mass advertising, which familiarized customers with the goods inside a store before they entered. The manual labor of self-service took place in stores, but name-brand marketing ushered in another newly pervasive form of mental labor: the information gathering or consumer research preceding the physical act of shopping. Much of shopping's authority was transferred from clerks to brands, which shoppers could use to identify reliable or desirable products on their own, without any further assistance once inside the store. Self-service was part and parcel of the rise of mass advertising and the transformation, as described by Stuart Ewen, of captains of industry into "captains of consciousness." During the creation of mass markets and the rise of a "consumer culture" in the U.S., innovations in self-service shopping helped push the limits of mass production and distribution.[52] Mass advertisements began not only to familiarize shoppers with merchandise before they went shopping, but ads for self-service itself also helped sell shoppers on the new labor waiting for them inside the stores. From the first survey of customer attitudes toward self-service supermarkets, shoppers have proven to be far more willing to take on the mental labor of shopping than its manual components.[53] It seems that deciding what to buy has always been more fun, and felt less like work, than the actual act of retrieving and purchasing something, and advertisers of self-service took advantage of this fact. The new consumer labor of self-service shopping, or parts of it anyway, could even pass for leisure.

Advertisements for self-service stores promoted a new way to provide shoppers with more control over the goods they selected, whether they came in knowing what they wanted or came to browse, which store owners preferred.[54] The new physical elements of shopping, and the absence of clerks to assist with them, were rationalized in presentations of a faster, more convenient way to shop for lower priced goods. When the Piggly Wiggly chain expanded beyond Memphis, its first national advertisements described self-service not only as "delightful" and "leisurely," but also as "sensible" and "health[y]" (because of advances in sanitary packaging). The Piggly Wiggly ads ran in magazines like the *Ladies' Home Journal*, where they presented self-service as "the quintessential example of U.S. values, personal control and free choice." In advertisements for Piggly Wiggly and other early self-service stores, "the additional work transformed into personal [as well as] financial benefits." Widespread acceptance of self-service was achieved, in part, by splitting its presentation in advertisements into mental and manual labor, more specifically into a series of decisions and tasks, which were in turn more fun and more efficient than the old way of shopping.[55] Promoters of self-service "all walked the same narrow path between rationalizing and romanticizing the [new] shopping experience."[56]

After "revolutionizing" the retail distribution of food in the U.S. during the 1930s and 1940s, self-service matured throughout the 1950s. First self-service expanded throughout the supermarket, from the grocery department

*Figure 1.5* "A nationwide vogue that leaves women free..."[57]

into other sections like meat and frozen foods. "[A]fter World War II, management aims were directed to 'complete' self-service in all food departments."[58] Whereas during the Great Depression consumers became more amenable to self-service and the lower food costs promised in exchange, labor shortages during World War further motivated merchants to make the change. At the end of the war, however, there were still only 28 stores in the U.S. that were completely self-service. Meat departments, for example, were still staffed by butchers. By 1953, though, over half of supermarkets "offered total self-service for prepackaged meats."[59] In many supermarkets, a butcher was still employed, but their workstations were increasingly

*Figure 1.6* Selling shoppers on the self-service "adventure."[60]

moved behind the scenes. Removing paid labor from the selling floor had two purposes: employees were less visible to customers, who were left with fewer options for assistance besides serving themselves, and the work of employees was further rationalized. "[S]elf-service markets utilize their help [*sic*] to the maximum. In many old-line stores, managers and employees frequently waste 25% of their time sitting around waiting for customers."[61] In the "old-line" stores, a clerk's job was to wait on (and for) customers. In self-service stores, the preparation of pre-packaged goods could take place behind the scenes, reducing the chance of (inefficient) interactions between employees and shoppers. The goods themselves should all be visible to shoppers, with as little back-room storage as possible, but the remaining work of paid employees – such as preparing meats and stocking shelves – should be invisible to shoppers.

While advances in packaging allowed meat departments to be converted to self-service, new methods of refrigeration meant that dairy sections could be converted as well, and "the increased popularity of frozen foods" after World War II allowed self-service to complete its colonization of the super-market. Refrigeration was also key beyond the supermarket. Along with the automobile, the home refrigerator allowed shoppers to purchase more food during each trip to the supermarket, contributing to the economies of scale at the heart of self-service. Many of the first – and most easily recognized – technologies facilitating self-service shopping were found in-store – turnstiles, shopping baskets, redesigned shelves, and floor plans. But self-service shop-ping was a sea change in retailing that could not have been accomplished solely inside stores. The rise, spread, and naturalization of self-service shop-ping was predicated on changing attitudes and expectations of American consumers, on mass production and mass consumption, and on the popular-ity of new consumer technology such as cars and refrigerators.

As self-service spread into all sections of the supermarket, supermar-kets came to dominate grocery shopping in the U.S. In 1955, *Life* magazine reported that, while only 5% of grocery stores nationwide qualified as super-markets, they accounted for over half of the retail food market, grossing over $17 billion.[62] Meanwhile self-service graduated from the supermarket and spread to other retail sectors. Filene's of Boston opened a self-service outlet store in its Needham, Massachusetts, warehouse in 1955, the first identifi-able self-service clothing store and the prototype for "discount department stores," which sprang up across the Northeast in the following years. Many of the supermarket's signature traits were exported to the discount outlets: a decentralized (and less expensive) location, huge floor space with mass displays of uniformly packaged goods, faster turnover of both inventory and personnel, lower-prices (and quality), long operating hours, liberal return policies and a parking lot several times larger than the store itself. By 1960, there were 140 self-service stores throughout New England selling "apparel and other soft goods on a low-margin, self-service basis," at a sales volume rate of $400 million per year.[63] Self-service also began to spread to depart-ment stores, further proof that the practice had taken root in the U.S. and was expanding vertically as well as horizontally throughout the retail landscape.

Just as supermarkets had first demonstrated profitability in the dense Northeast and then spread across the U.S., it was estimated that nation-wide over 400 new "discount department stores" would open in 1961 alone, following in Filene's footsteps and raising the self-service soft goods sales volume to over $3 billion annually. And just as significant as these skyrock-eting projections were their sources. The 1961 estimates were delivered by Nathaniel Schwartz, the first editor of *Discount Merchandiser*, during his keynote address at the first Conference on Discount Department Stores and

Discount Food Store Operations, held at the University of Massachusetts, Amherst, in April 1961. In June 1960 the trade paper *Modern Retailer* was launched, and the following May *Discount Merchandiser* began publication as a monthly journal. In the wake of self-service's expansion from supermarkets into apparel and other soft goods sectors, these new periodicals and conferences collected the threads of research and analysis devoted to self-service, which had been spinning out of retail and marketing journals for decades, and solidified self-service studies as a discrete field of business scholarship.[64]

## Employing the Customer

In the most basic terms, self-service indicates that employees' paid labor has become consumer labor. The same individual might perform the same task; for instance, a grocery clerk laid off during the Great Depression still presumably shopped for groceries and conceivably still pulled the same items down from the same shelves. But whereas the retrieval of goods had been performed for customers and employers, in exchange for a wage, now it was done for oneself. Because consumers began to take possession of merchandise before paying for it, the legal status of goods during this new interval came into question, and the adjudication of shopping in self-service stores challenged established law regarding the contract of sale. What were the legal responsibilities of buyer and seller during this ambiguous new interval? Did possession equal purchase, even before money changed hands? In the end, not only was much of the labor of shopping taken over by customers inside self-service stores, but shoppers also assumed legal responsibility for goods before purchasing them. Self-service shopping required an update to the contractual mechanics of retail transactions.

There were also new dangers introduced by mass display, and it fell to the courts to determine whether, when and how the manufacturer or retailer (or neither) was responsible for a shopper's safety in a self-service store. The first lawsuits brought against self-service stores by shoppers reformed the liability of their commercial relationship. Some legal responsibilities of employers toward their employees were extended to consumers, and shoppers were recognized as legally entitled to some of the same health and safety protections extended to workers in the store. The courts deduced that since shoppers in a self-service store were assigned the tasks of a clerk, they also deserved some of the rights and protections of employees. Of course this quasi-employee status cut both ways, and self-service customers also became culpable in the store in new ways. The "doctrine of assumed risk," which had historically protected employers from liability for employees' injuries on the job, was extended for the first time to consumers. Merchants were insulated from liability for injuries sustained by shoppers helping themselves. Self-service reinforced the conventional wisdom that possession is nine-tenths of the law, but it also upturned the age-old retail maxim about the customer always being right.

Self-service challenged the "well-established sales rule that the cash sale occurs at the time of payment."[65] One key early case directly took on the question of whether "the sale concept could be extended to include the necessary preliminary components" of self-service, namely the customers' possession of merchandise before paying for it. In *Laskey v. Economy Stores*, the plaintiff was injured when a bottle of tonic exploded, after she had selected it herself and taken possession of it, but before she had paid for it. The store was found not liable, but a subsequent review article (the first to review self-service case law) suggested the lawsuit in this instance had been filed against the wrong party. The reviewer concluded the "logical remedy" in such a case was a suit against the manufacturer for "breach of warranty" rather than against the retailer. If the store could not be found liable for such "merchandizing hazards," then a lawsuit against the manufacturer might motivate other makers of potentially dangerous products to insure themselves against liability. The upshot would be some "protection of the comparatively more vulnerable consumer." The first law review article devoted to self-service recommended transferring liability for a shopper's safety from the distributor of goods to the producer.[66]

What were the competing interests to consider in self-service lawsuits? Unlike the aforementioned review article, judges in all early self-service cases disregarded the interests of consumers and employees as general categories, and "confined [themselves] to considerations of the relations of the individual interests of shopkeeper and customer."[67] These "individual interests" were defined as control of the merchandise vs. freedom of choice, respectively, and it was the interval between possession and purchase when they came into conflict. Consequently, the self-service sale was broken down into components, or phases. The mass display of goods for sale in self-service shops constituted "invitations to bargain" proffered by the merchant. But selecting a product and carrying it around the store could not unequivocally or irredeemably lead to possession; otherwise, shoppers could never change their mind and replace an item before purchasing it, which "would quickly destroy the popularity of [self-service] shops."

In this new merchant-shopper relationship the traditional roles of buyer and seller were rearranged. In a self-service store, who made the offer to exchange goods for currency, and who accepted? Was the display of goods with set prices an offer to sell, or, was it the customer who made an offer to buy, by selecting a product and bringing it to the counter? Since it remained the case that "there is no sale effected until the buyer's offer to buy is accepted" by the merchant, the new interval of possession-before-purchase was deemed "the stage of preliminary negotiation."[68] A role reversal – whereby the buyer made an offer and the seller accepted – allowed the merchant to retain control until the point of sale, while responsibility could be transferred to the shopper at the earlier moment of possession. Self-service meant consumers legally assumed responsibility for a product upon touching it. Merchants, however, retained control over their merchandise until it was purchased.

The first lawsuit against a self-service supermarket was filed in 1938, in Shelby County, Tennessee (also home to Clarence Saunders' first Piggly Wiggly store in Memphis). A negligence suit was brought by executors to the will of a woman who fell over a shopping basket left on the floor in a Kroger Grocery store and later died from sustained injuries. The circuit court found in favor of the defendant, Kroger, and the state court of appeals upheld the lower decision.[69] Pertinent facts in the appeal hearing included the shopping basket itself, which undisputedly belonged to Kroger as a "necessary part of the equipment of a self-service store." The woman who left the fatal basket in the aisle was identified, and the facts of the case stated she had left her basket unattended for between twenty and thirty minutes in an isle near the "cake counter," while she shopped and visited at the meat counter in the back of the store. It was also established that it was common practice to leave one's basket unattended on the floor in this particular store, and that as a regular shopper there for four years, the deceased was familiar with the custom. The deciding factor in the case was that no employee had been made aware of the fateful basket, and so Kroger, the employer, could not be found liable for it. The accident happened on a Saturday evening shortly before Christmas, and there were only three employees on hand at the time, "who were very busy." Ironically, this understaffing helped insulate the store from negligence charges. The court denied the charge of negligence brought against the store on the grounds that the woman who fell and died was guilty of "contributory negligence." In other words, she should have been more careful, on the look out for unattended baskets and other hazards. She "assumed any risk incident to shopping in the store," including the risks introduced by self-service. The doctrine of assumed risk was categorically expanded from its historical application to employees injured or causing injury while on the job. Here it was being applied for the first time to customers.

The second pivotal self-service case also applied the employer-employee relationship to a store and a shopper, as it applied specifically to the dangers of mass display. Again employee status was extended to a consumer, but in this instance to rule for the plaintiff. In 1945, a woman sued her local tea shop for negligence after sustaining injuries from a can falling from a shelf onto her head. The Supreme Court of New York State upheld the judgment of a Manhattan Municipal Court finding the store to have been negligent.[70] The NYS Supreme Court found that self-service stores "delegated to its customers the right to handle merchandise freely and had invited them on the premises; thus, the store has a foreseeable duty to guard ... the customers," no less than its employees, from injury. The self-service store was responsible to its customers in new ways, the court ruled, because it was asking them to take on new tasks. These tasks were of course not new, but rather newly assigned, and the court found in the employee-employer relationship between clerk and storeowner precedence to rule in *Robinson*. The self-service patron acted in a "dual capacity ... as a customer, and, also, as a store service clerk." A self-service store conscripted its customers in this sense as employees, "their agents or servants" assigned to "move, take, remove and replace articles on the shelves." Consequently, the court ruled a proprietor of a self-service store "should be held as much responsible for their acts of carelessness and negligence and for any resulting injury as in the case of an actual employee."

Perhaps it is not surprising that an appeals court in the Tennessee county identifiable as the birthplace of self-service would side with the merchant, while in relatively liberal and labor-friendly New York, a state supreme court judge would draw on employee protections to rule for the consumer. However, there is a fundamental difference in liability between injuries sustained from an unattended basket and a falling can. The unattended basket was full of goods that had already been selected and pulled from the shelves, and the store was found not liable for the basket because a shopper had already taken possession of the goods inside it. In *Robinson*, on the other hand, the plantiff had not selected the falling can. The doctrine of "assumed risk" applied to shoppers for risks introduced by other shoppers in a self-service store, such as an unattended basket, but not for risks introduced simply by being in the store, such as falling cans. *Robinson v. A&P Tea* was more straightforward than *Gargaro v. Kroger* because it did not involve the new retail interval between possession and purchase introduced by self-service shopping.

These cases legally transferred much of the responsibility of self-service shopping to customers, once retailers began implementing self-service as a new type of store organized around mass marketing and mass display. The new interval between possession and purchase required revisions to the technical relationship between buyers and sellers. Legally, sellers retained control during the new interval, but buyers took on more responsibility sooner. In this sense, the law aided and abetted the naturalization of self-service shopping as a technological assemblage in which consumers took on more liability as well as labor.

## Ring It Up

Self-service shopping diffused disemployment, whereby many customers took up the slack for each expendable clerk. However, it is incomplete to conclude that the formative "work transfer" of self-service shopping was simply or exclusively a transfer of tasks and responsibilities from clerks to customers. While the store clerk was deskilled with the rise of self-service, not all of the clerk's functions were transferred to shoppers. Self-service entailed a new division of labor, whereby much of the clerk's work was transferred to customers. The transfer of a clerk's work, however, proceeded on two tracks; the clerk's tasks were fragmented and reassigned to not one, but two workers within the store. One is the customer, and the other is the cashier. The skilled position of sales clerk was divided into two new things: the unpaid work of self-service and the unskilled job of cashier. Nona Glazer has used census data to point out "the emergence of the 'cashier' as a distinct occupation" during the 1940s, and this emergence followed directly on the heels of self-service becoming the standard mode of grocery shopping. The new division of labor within self-service stores meant that, as Paul du Gay has put it, "'deskilling' for the shop worker became a form of 'enskilling' for the customer." In self-service stores paid labor was "externalized" into unpaid work, and simultaneously skilled labor was "degraded" along the lines of Harry Braverman's influential critique of automation in manufacturing.[71]

In her use of census data to analyze a two-track work transfer within self-service retail, Glazer demonstrates that both of these transfers were thoroughly gendered.[72] Furthermore, it is not simply the case that each transfer is gendered; rather, they work in tandem. It is not just that women were hired for lower-paid, lower-skilled positions or that externalized labor flows from employed men to self-serving women. Both are true, in that the skilled jobs of male clerks tended to be converted into unskilled jobs for female cashiers, while self-service shoppers, also overwhelmingly female, compensated for the "degradation" of retail sales work. But self-service entailed a simultaneous combination of deskilling and enskilling. Self-service "brought women into *paid* employment at the same time women consumers were increasingly asked to do self-service." The new division of labor within retail meant "women shoppers do *some* of the work that men once did as sales clerks [while] women paid workers do *some* of that work as cashiers." The skilled sales clerk positions were almost all held by men, and with the rise of self-service, "women did parts of what had been a single job." Self-service not only entails the transfer of work, it also provides the occasion to fragment it, into unskilled labor as well as consumer labor. The transactional element of shopping changed when self-service split possession from purchase. The transactions themselves, however, did not (yet) become self-service. Rather, the skilled labor of clerks was fragmented into not only new tasks and responsibilities for shoppers, but also into newly deskilled labor – namely, cashiers. "Cashier" did not become a recognizable category of employment in the U.S. when store clerks and other employees began using cash registers as part of their job, which happened during the 1880s. The job title did not emerge until sixty years later, on the heels of self-service shopping, when some employees were left doing nothing else on the job *but* operating a cash register.

The cash register as well as the cashier became crucial technology in the self-service shopping assemblage. Founded in 1884 in Dayton, Ohio, National Cash Register (NCR) was the first company in the world to manufacture cash registers. (A century later, NCR would become a leading manufacturer of ATMs.) The cash register itself was invented by James Jacob Ritty, a saloon keeper in Dayton, to stop employees from stealing his profits. Legend has it that Ritty was "so troubled by the petty thievery which was ruining his business that he suffered a breakdown."[73] A coal merchant in Dayton, John Patterson ordered Ritty's first two machines after noticing some of his hired clerks were not recording every sale. The cash registers increased Patterson's profits so dramatically that he invested in NCR and quickly became a board member with big ideas about the company's future.

NCR looms large in the annals of American business history, and the cash register itself is arguably only the third most significant contribution made by Patterson's company to the development of modern commercial practices in the U.S. First of all, NCR's competitive prowess has served as a case study for business analysts of "a monopoly able to consolidate and defend its dominant position by means of an effective predatory strategy."[74] This strategy included the first "knock-out fund," announced in an 1890 issue of the company newsletter, which announced to competitors as well as employees: "We propose hereafter to set aside, say, $5 on each register made, for a knockout expense fund to be devoted to maintain a monopoly."[75] During the 1890s, NCR "knocked out" virtually all of its competitors. Traces of more than eighty companies developing variations on "Ritty's Incorruptible Cashier," as it was known at the time, have been found, but by 1897, NCR had reduced its number of known competitors to three. NCR maintained its dominance even after antitrust lawsuits, and at the onset of World War I controlled ninety-five percent of the cash register market.[76]

*Figure 1.7* Ritty's "incorruptible cashier."[77]

NCR's notoriety as a "predatory monopoly," however, did not mitigate the impact of Patterson's innovative sales methods, which are arguably the company's most lasting legacy. Daniel Boorstin, in his sweeping social history *The Americans*, credits Patterson with "develop[ing] salesmanship itself into a new institution," and his impact in this capacity is difficult to overestimate. Patterson developed a system for managing a corporate sales staff still widely used today: divide up a company's clients (and potential clients) into geographic territories to eliminate competition among the sales "team," and then institute a quota system to motivate each member of the team. The establishment of individual quotas for each NCR salesman (exclusively) "took the guess work out of selling," and put it instead "on a firm mathematical foundation so that production could confidently be scheduled."[78] During the 1950s, the *Saturday Evening Post* presented a history of the cash register that compared Patterson's "amazing formula for getting rich" to other American business titans. "Patterson's theory of business," for instance, was described as "the opposite of Henry Ford's." As demand for cash registers increased, manufacturing costs decreased, yet Patterson kept prices high. "Instead of cutting the price to increase the demand, as Ford did, Patterson created demand through salesmanship, and distributed the profits liberally to workmen, salesmen, executives and himself." Bucking not only economies of scale but also the conventional wisdom of his day, Patterson also insisted that great salesmen were made, not born, and he instituted the first systematic training of salespeople, complete with a 450-page primer that sales representatives had to memorize before embarking on a seven-week training course, held at a farm near Patterson's home town near Dayton, known (for some reason!) as NCR's "Sugar Camp."[79]

One of the drastic changes Patterson introduced into sales was the replacement of pressure tactics as the profession's stock-in-trade with comprehensive knowledge of the product for sale; however, Patterson still employed catchy slogans to impress NCR's potential customers. Sales pitches included warnings to employers about "cash drawer weakness" among their employees, and NCR advertisements turned the cash register into a monument of financial security: "You insure your life. Why not insure your money too! A National cash register will do it." NCR salespeople were also often purchased new wardrobes, at the company's expense, before being turned loose on clients, and prohibitions were issued against chewing gum or tobacco, sitting before the potential client is seated, and wasting time with personal stories. Patterson began scientifically managing his company's male, white-collar labor, his salesmen, more than a decade before (as the next chapter describes) telephone operators became "unrivaled" in the degree of scientific management visited upon them.[80] NCR began practicing this white-collar version of scientific management even before Taylor's infamous tome about the factory. Patterson's pioneering techniques of "intensive supervision," as he called it, included "the right to regulate a salesman's personal life as well as his ... selling."[81] NCR under Patterson's control developed "a costly program of employee welfare and entertainment" that eclipsed the coddling of operators by "Ma Bell" during the same era. A job with NCR included "company recreation rooms ... banquets and picnics ... courses, concerts and lectures in the company auditorium, [and] he land-scaped the factory grounds into an 'industrial garden.'" Patterson may have conceived of the relationship between price and demand in opposition to Henry Ford, but Patterson "exhorted employees to morality, health and patriotism" no less than Ford did. "It pays" even became a tag line of sorts for Patterson, used to explain his unprecedented largesse. NCR anticipated scientific management, but Patterson was also a pioneer of the "humane" corporate workplace.[82]

Boorstin includes his chapter about Patterson and NCR in a section of *The Americans* titled "Statistical Communities," and ultimately he finds in the cash register origins of "an American system of merchandising and accounting as distinctive as the American system of manufacturing." Self-service became the distinctively American system of distribution. "Along with the cash register came the calculating machine," and Boorstin credits these two technologies, alongside Patterson's business practices, with cat-alyzing "the nation's number consciousness." Patterson and NCR shifted American businesses away from qualitative assessments in favor of statis-tical analyses. Until the cash register, "it was the rare merchant who knew the precise amount of his income." Once cash registers and adding machines became commonplace, Boorstin concludes, "businessmen and consumers could not help thinking quantitatively." Boostin's suggestion that the cash register helped quantify measures of success and failure at the sales till is persuasive, and Patterson pioneered statistical sales "metrics" at NCR that are still widely used today.[83]

In addition to strenuous promotion of the cash register and unprece-dented methods of selling them, Patterson continued to improve on the orig-inal design once he assumed control of NCR, for instance adding paper rolls to provide customers as well as the company with a printed receipt for each transaction and designing the first electrical cash registers as early as 1906. NCR cash registers also came with bells affixed to them, which rang every time the drawer was opened. The bells help storeowners police their clerks, but an audible cash register also imbued "the shopkeeper's smallest transac-tion [with] a new publicity. ... Shopping now was a semipublic, communal activity, announced by the ringing of bells."[84] Whereas telephones began ringing inside people's homes and offices, cash registers were designed to ring in commercial spaces. The cash register did not ring to prompt anyone to begin using a machine, like a ringing telephone (or dial tone) did. A cash register rang so that everyone *else* in the store, besides the paying customers and the clerk serving them, would know that a sale was being completed. The phrase "ring it up" dates to the advent of the cash register, and just as "dialing" continues to pass as a label for the process of establishing a tele-phone connection, "ring it up" continues to indicate that a sale has been completed, regardless of what type of noise a particular cash register may make, if any, and even when a cash register is nowhere in sight. (An onomatopoetic descendent, "cha-ching," has also come to signify success more generally.)

The introduction of cash registers met with some resistance during the late nineteenth century, organized by clerks and other keepers of the till into "Protective Associations" to oppose the new machines.[85] Concerns about the new machines were not couched in terms of job loss or deskilling, and organized resistance to cash registers considered the machines to be more an insult than a replacement. When employees had the cash register explained to them, they "resented it as a slur on their character."[86] (Even though Ritty, Patterson and other early denizens of the cash register all claimed to see spikes in their profits once the machines were installed, observers like Boorstin fail to mention that employees may have also recognized in "Ritty's Incorruptible Cashier" a threat to pilfering, which was in any number of 19th century occupations an established means of informally supplementing one's income.) The subsequent deskilling of clerks into cashiers met with virtually no organized resistance, and in fact some labor advocates even welcomed self-service into grocery stores and other retail outlets, on the assumption that cashiers would be less-satisfied workers than clerks had been, and therefore more receptive to unionization.[87]

## Conclusion: Checking Out

Grocery stores were at the vanguard of self-service, and supermarkets are where the practice took hold. Self-service has long been the norm for grocery shopping, and during the first years of the twenty-first century supermarkets extended self-service from the selling floor to the till, or from "shopping as choosing" to "shopping as making a purchase."[88] In contemporary supermarkets, self-check out comprises a new technological assemblage, and "we have to consider the effectivity of a whole array of machines, practices, habits, attitudes, ideas" before we can understand how scanning and bagging one's own groceries and swiping a debit or credit card to pay for them can become second nature for so many shoppers.[89] Any assemblage absorbs elements from others; for self-check out, this includes scales and scanners along with card swipes and keypads. The automation of cashiers' labor also spawned the Universal Product Code, as well as scanners for it, which shoppers now utilize in self-check out lanes. The history of self-service shopping reaches at least as far back as shopping baskets and Piggly Wiggly's patented turnstiles; however, the pairing of "self-service" and "technology" is a much more recent phenomenon. One of the first systematic studies of self-service technologies [SSTs] was published in 2000, in the *Journal of Marketing*, the same journal that in 1941 published Zimmerman's manifesto, "The Supermarket and the Changing Retail Structure." The more recent article concludes that "[w]ith SSTs, customers create the service for themselves, so [they] accept more of the responsibility for the outcome" than in traditional service interactions with paid (blamable) employees.[90] The authors note "important managerial implications" of this technological transfer of responsibility from employees to consumers, such as: "if customers accept partial responsibility in dissatisfying situations, they may be more likely to use the SST in the future." This chapter has narrated a history of the transfer of shopping's responsibilities from employees to consumers. The emergence and acceptance of self-service shopping involved legal as well as spatial, technological and managerial reorganizations. The initial self-service technologies included new ideas as well as new machines and new representations as well as new rules. No conspiracy was necessary; by constructing new spaces and passing new legislation, retailers and policy makers acted in concert to interpolate American consumers as self-serving subjects.

In 2005 over $600 billion worth of retail transactions took place in self-check out aisles, and the market has increased by 10% each year since then.[91] During fall, 2011 a spate of news articles described among shoppers a backlash of sorts against self-service check out in supermarkets.[92] Supermarket spokespeople claimed their companies were responding to customer dissatisfaction by removing SCO assemblages in favor of more human cashiers. The punch line is that changes in barcodes and, especially, increased transactional activity via smart phones are rendering obsolete the (still relatively new) SCO assemblages. The overall figures for self-check out continue to rise, but grocers unplugging their SCO assemblages are doing so in anticipation of upgrading their self-payment assemblages. Near-field communication [NFC] chips, for instance, enables shoppers to use smart phones for scanning items and paying for them. New technological capacities and changing attitudes about them "undermine any assemblage's stability. ... Agency active in any assemblage isn't necessarily *human* agency, which is only "possible or not possible depending on the particular assemblage."[93] The untimely demise of SCO assemblages does not mean that shoppers opposed to them have exercised any agential resistance. Rather, self-payment is migrating from technologies of consumer labor inside stores (and outside airports, etc.) to ones that shoppers carry with them. Self-service shopping has always lowered labor costs for retail merchants by reassigning service work to customers using machines. The use of personal technology within self-payment assemblages also offloads purchasing and maintenance costs onto shoppers.

The most productive self-service technology of all remains the self-serving shopper. This designation is difficult, if not impossible, to quantify in part because the productivity of self-serving consumers is ethical as well as economic, a matter of expectations as well as efficiency. It should be clear, however, that the greatest return on investment in self-service historically has been not reduced labor costs. Those savings have been substantial, and observers and analysts often pinpoint them as the impetus for implementing self-service wherever possible. In the early days, these reduced costs were touted in advertisements as being passed on to shoppers as lower prices, savings earned in exchange for a little more work. Ultimately, however, the success of self-service shopping – M. M. Zimmerman's "revolution in distribution" – cannot be measured in lower labor costs for employers or lower prices for shoppers. The success story of self-service is a story of historical change wrought not only in consumers' actions, but in their attitudes as well. Whether or not shoppers welcome self-service or resist it, and historically they have done both, the bottom line is that most have grown to expect it.

## Notes

1. *Double Indemnity*, directed by Billy Wilder (Los Angeles: Paramount, 1944).
2. Strom, *Beyond the Typewriter: Gender, Class and the Origins of Modern American Office Work, 1900–1930* (U Illinois P, 1992): 173–4.
3. See, for example, Susan Porter Benson, *Counter Cultures: Saleswomen, Managers and Customers in American Department Stores, 1890–1940* (Urbana, IL: University of Illinois Press, 1984).
4. Paul du Gay, "'Numbers and Souls:' Retailing and the De-Differentiation of Economy and Culture," *The British Journal of Sociology* 44 (1993): 572.
5. Sharon Zukin, *Point of Purchase: How Shopping Changed American Culture* (New York: Routledge, 2004): 72.
6. Rachel Bowlby, *Carried Away: The Invention of Modern Shopping* (London: Faber and Faber, 2000): 31.
7. On the rise of Woolworth's, see John P. Nichols, *Skyline Queen and the Merchant Prince* (New York: Trident Press, 1973).
8. Bowlby, 8.
9. James Beniger, *The Control Revolution: Technological and Economic Origins of the Information* Society (Cambridge, MA: Harvard University Press, 1986): 333.
10. Kerry Segrave, *Vending Machines: An American Social History* (Jefferson, NC: McFarland, 2003): 3.
11. Ibid., 21. Nearly a century later, Nicholas Carr and Craig Lambert would rehash these ideas for digital technology and computerized robotics. See the introduction, pp. 16–18.
12. Paul H. Nystrom, *Retail Selling and Store Management* (New York: Appleton and Co., Inc., 1925), quoted in Bowlby, 31.
13. Bowlby, 31.
14. Walter W. Hurd, "Ultra self-service," *Billboard*, June 8, 1946, p. 100, quoted in Segrave, 121.
15. Segrave, 13; Beniger, 333.
16. Ibid.
17. M. M. Zimmerman, *The Supermarket: A Revolution in Distribution* (New York: McGraw-Hill, 1955), 24.
18. Kent Lutey, "Lutey's Marketeria–Self-Service Grocers," *Montana* 28 (1978): 50–7. See also Nona Glazer, *Women's Paid and Unpaid Labor: The Work Transfer in Health Care and Retailing* (Philadelphia: Temple University Press, 1993), 51–2.
19. William J. Regan, "Self-service in Retailing," *Journal of Marketing* 24 (1960): 43.

20. Images reproduced from James M. Mayo, *The American Grocery Store: The Business Evolution of an Architectural Space* (Westport, NC: Greenwood Press, 1993), 91; and Susan Strasser, *Satisfaction Guaranteed: The Making of the American Mass Market* (New York: Pantheon Books, 1989), 250.
21. Zimmerman, 22.
22. Carl W. Dipman and John E. O'Brien, *Self-Service and Semi-Self-Service Food Stores* (New York: The Butterick Co., Inc., 1940), 35.
23. Zukin, 72.
24. Glazer, 53.
25. Beniger, 333–4.
26. Slack and Wise, 131.
27. Zukin, 71.
28. Carl W. Dipman and John E. O'Brien, eds., *Self-Service and Semi-Self-Service Food Stores* (The Butterick Co., Inc., 1940), 7.
29. Dipman and O'Brien were the editor and associate editor, respectively, of *The Progressive Grocer*.
30. Ibid., 14.
31. Ibid., 70.
32. Ibid., 86.
33. Ibid., 42.
34. Bowlby, 174.
35. Strasser, 31.
36. Ibid., 32.
37. In 1900 ... only 7 percent of ... packaged goods pictured their product; by 1925, 35 percent did. Beniger, 337.
38. Strasser, 32.
39. Beringer, 337.
40. Bowlby, 88.
41. M. M. Zimmerman, "The Supermarket and the Changing Retail Structure," *Journal of Marketing* 5 (1941), 405.
42. Mayo, 153.
43. Bowlby, 153.
44. Zimmerman, *The Supermarket: A Revolution in Distribution* (New York: McGraw-Hill, 1955).
45. Zimmerman, "The Supermarket and the Changing Retail Structure," 405.
46. Zimmerman, *The Supermarket*, 17–18, emphasis added.
47. Mayo, 146.
48. Statistics and quotations in this paragraph from Zimmerman, "The Supermarket and the Changing Retail Structure," p. 402–3.
49. Zimmerman, *The Supermarket*, 30.
50. Mayo, 117–18.
51. Ibid., 153.
52. Stuart Ewen, *Captains of Consciousness: Advertising and the Social Roots of the Consumer Culture* (New York: McGraw-Hill, 1976).
53. Myrtle M. Lohner, "Customer Attitude Toward Chicago Grocery-Store Practices," *The Journal of Business of the University of Chicago* 10 (1937): 233–50.
54. In the early 1960s, business analysts noticed "an increasing number of consumer purchases [we]re being made without advance planning," and they included self-service and mass advertising, alongside mass display and low prices in the

"impulse mix." See Hawkins Stern, "The Significance of Impulse Buying Today," *Journal of Marketing* 26 (1962), 59–62.

55. Glazer, *Woman's Paid and Unpaid Work,* 99–100.
56. Zukin, 78.
57. Advertising Archive, Department of Media, Culture and Communication, NYU. Accessed February 29, 2016.
58. Mayo, 177.
59. Ibid.
60. Ibid.
61. Dipman and O'Brien, 10.
62. "Shopper's Delight," *Life* magazine, January 3, 1955, 38.
63. Statistics and quotations in this paragraph from George Tallman and Bruce Blomstrom, "Self-Service in Soft Goods," *Journal of Marketing* 25 (1961): 133–43. Tallman and Blomstrom critiqued the name "discount department stores," arguing that it was an attempt by anxious investors to "capitalize on the already established public association of 'discounts' with unusual value, and of 'department store' with an attractive quality and breadth of merchandise." A more precise, if less appealing name for the new genre of retail outlet, they suggest, would be "soft-goods, self-service supermarkets."

    Sellers of "heavy goods" resisted the self-service format. Retailers and analysts alike assumed the size and weight of products like appliances and automobiles prohibited transferring to consumers the manual labor of purchasing them. Although the delivery of heavy goods still requires manual labor in the form of pick-up or delivery, telecommunications technologies subsequently have been utilized as tools for shoppers to assume much of the mental labor of shopping for heavy as well as soft goods, namely consumer research and comparison shopping, as well as selection, purchase and payment online.
64. Dates and statistics from William J. Regan, "Full Cycle for Self-Service?" *Journal of Marketing* 25 (1961): 7.
65. *Lasky v. Economy Stores,* 65 N.E. (2nd) 305 (Mass. 1946).
66. "Self-service Merchandising and Consumer Protection before Sale Consummated (Lasky v. Economy Stores, Mass. 1946), *Columbia Law Review* 47 (1947), 156–8.
67. J. Unger, "Self-Service Shops and the Law of Contract," *The Modern Law Review* 16 (1953): 369–70. All quotations this paragraph.
68. J. J. Montrose, "The Contract of Sale in Self-Service Shops," *The American Journal of Comparative Law* 4 (1955): 238–9.
69. *Gargaro v. Kroger Grocery & Baking Co.* 22 Tenn. App. 70l 118 S.W. 2d 561; 1938.
70. *Robinson v. A&P Tea, Co., Inc.* 184 Misc. 571; 54 N.Y.S. 2d 42; 1945.
71. Nona Y. Glazer, "Servants to Capital: Unpaid Domestic Labor and Paid Work," *Review of Radical Political Economics* 16 (1984): 74; du Gay, "'Numbers and Souls,' 572; Ursula Huws, *The Making of a Cybertariat: Virtual Work in a Real World* (New York: Monthly Review Press, 2003): 179, 23; Harry Braverman, *Labor and Monopoly Capitalism: The Degradation of Work in the Twentieth Century* (New York: Monthly Review Press, 1974). Hews uses the phrase "externalizations of labor" to describe the processes by which paid service labor becomes what she calls "unpaid consumption work."
72. Glazer, "Servants to Capital," p. 76, emphases in original. All quotations in this paragraph.

73. Daniel Boorstin, *The Americans: The Democratic Experience* (New York: Vintage, 1974): 200–1.
74. Ibid.
75. *The N.C.R.*, May 22, 1890.
76. "NCR Corporation," *International Directory of Company Histories*, Vol. 30 (St. James Press, 2000).
77. Image from the Ohio History Central webpage for James Ritty. Accessed March 14, 20016. http://www.ohiohistorycentral.org/w/James_Ritty.
78. Boorstin, p. 202.
79. Charles Wertenbaker, "Patterson's Marvelous Money Box," *Saturday Evening Post*, October 3, 1953.
80. Steven H. Norwood, *Labor's Flaming Youth: Telephone Operators and Worker Militancy, 1878–1923* (U Illinois P, 1990): 1.
81. Wertenbaker, "Patterson's Marvelous Money Box."
82. Boorstin, 202–3; Andrew Ross, *No-Collar: The Humane Workplace and Its Hidden Costs* (New York: Basic Books, 2003).
83. Boorstin, 200, 203, 205.
84. Ibid., 201.
85. Samuel Crowther, *John H. Patterson : Pioneer in Industrial Welfare* (Garden City, N.Y.: Garden City Publishing Co., 1926), 92.
86. Boorstin, 202.
87. For an example of this sanguine approach to the deskilling of clerks into cashiers, see Michael Harrington, *The Retail Clerks* (New York: Wiley and Sons, 1962).
88. Bowlby, *Carried Away*, 31.
89. Slack and Wise, 158.
90. Matthew L. Meuter, Amy L. Ostrom, Robert I. Roundtree, May Jo Bitner, "Self-Service Technologies: Understanding Consumer Satisfaction with Technology-Based Service Encounters," *Journal of Marketing* 64 (2000): 50–64.
91. Greg Buzek, President IHL technology research firm, "Self-Service Checkout is Alive and Well," Nov. 15, 2011. Accessed February, 29, 2016. http://www. slideshare.net/G3Com/selfcheckout-is-alive-and-well.
92. See, for example, Stephanie Reitz, "Shoppers mixed about bagging self-checkout," *Associated Press*, September 26, 2011; James Joyner, "Supermarket Self-Checkouts Being Replaced with People," *Outside the Beltway*, September 27, 2011.
93. Slack and Wise, 158.

## Bibliography

Beniger, James. *The Control Revolution: Technological and Economic Origins of the Information Society.* Cambridge, MA: Harvard University Press, 1986.

Benson, Susan Porter. *Counter Cultures: Saleswomen, Managers and Customers in American Department Stores, 1890–1940.* Urbana, IL: University of Illinois Press, 1984.

Boorstin, Daniel. *The Americans: The Democratic Experience.* New York: Vintage, 1974.

Bowlby, Rachel. *Carried Away: The Invention of Modern Shopping.* London: Faber and Faber, 2000.

Braverman, Harry. *Labor and Monopoly Capitalism: The Degradation of Work in the Twentieth Century*. New York: Monthly Review Press, 1974.

Crowther, Samuel. *John H. Patterson: Pioneer in Industrial Welfare*. Garden City, N.Y.: Garden City Publishing Co., 1926.

Dipman, Carl W. and John E. O'Brien. *Self-Service and Semi-Self-Service Food Stores*. New York: The Butterick Co., Inc., 1940.

*Double Indemnity*. Directed by Billy Wilder. Los Angeles: Paramount, 1944.

Ewen, Stuart. *Captains of Consciousness: Advertising and the Social Roots of the Consumer Culture*. New York: McGraw-Hill, 1976.

*Gargaro v. Kroger Grocery & Baking Co.* 22 Tenn. App. 70l 118 S.W. 2d 561; 1938.

Gay, Paul du. "'Numbers and Souls:' Retailing and the De-Differentiation of Economy and Culture." *The British Journal of Sociology* 44 (1993): 563–87.

Glazer, Nona Y. "Servants to Capital: Unpaid Domestic Labor and Paid Work." *Review of Radical Political Economics* 16 (1984): 61–87.

Glazer, Nona. *Women's Paid and Unpaid Labor: The Work Transfer in Health Care and Retailing*. Philadelphia: Temple University Press, 1993.

Harrington, Michael. *The Retail Clerks*. New York: Wiley and Sons, 1962.

Hurd, Walter W. "Ultra self-service," *Billboard*, June 8, 1946.

Huws, Ursula. *The Making of a Cybertariat: Virtual Work in a Real World*. New York: Monthly Review Press, 2003.

Joyner, James. "Supermarket Self-Checkouts Being Replaced with People." *Outside the Beltway*, September 27, 2011.

*Lasky v. Economy Stores*, 65 N.E. (2nd) 305 (Mass. 1946).

Lohner, Myrtle M. "Customer Attitude Toward Chicago Grocery-Store Practices." *The Journal of Business of the University of Chicago* 10 (1937): 233–50.

Lutey, Kent. "Lutey's Marketeria–Self-Service Grocers." *Montana* 28 (1978): 50–7.

Mayo, James M. *The American Grocery Store: The Business Evolution of an Architectural Space*. Westport, NC: Greenwood Press, 1993.

Meuter, Matthew L., Amy L. Ostrom, Robert I. Roundtree, and May Jo Bitner. "Self-Service Technologies: Understanding Consumer Satisfaction with Technology-Based Service Encounters." *Journal of Marketing* 64 (2000): 50–64.

Montrose, J. J. "The Contract of Sale in Self-Service Shops." *The American Journal of Comparative Law* 4 (1955): 238–9.

Nichols, John P. *Skyline Queen and the Merchant Prince*. New York: Trident Press, 1973.

Norwood, Steven H. *Labor's Flaming Youth: Telephone Operators and Worker Militancy, 1878–1923*. Urbana, IL: University of Illinois Press, 1990.

Nystrom, Paul H. *Retail Selling and Store Management*. New York: Appleton and Co., Inc., 1925.

Regan, William J. "Self-Service in Retailing." *Journal of Marketing* 24 (1960): 43.

Regan, William J. "Full Cycle for Self-Service?" *Journal of Marketing* 25 (1961): 15.

Reitz, Stephanie, "Shoppers mixed about bagging self-checkout," *Associated Press*, Septemeber 26, 2011.

*Robinson v. A&P Tea, Co., Inc.* 184 Misc. 571; 54 N.Y.S. 2d 42; 1945.

Ross, Andrew. *No-Collar: The Humane Workplace and Its Hidden Costs*. New York: Basic Books, 2003.

Segrave, Kerry. *Vending Machines: An American Social History*. Jefferson, NC: McFarland, 2003.

"Self-service Merchandising and Consumer Protection before Sale Consummated (Lasky v. Economy Stores, Mass. 1946)." *Columbia Law Review* 47 (1947): 156–8.

Slack, Jennifer Daryl and J. MacGregor Wise. *Technology and Culture: A Primer*, 2nd edition. New York: Peter Lang, 2015.

Stern, Hawkins. "The Significance of Impulse Buying Today." *Journal of Marketing* 26 (1962): 59–62.

Strasser, Susan. *Satisfaction Guaranteed: The Making of the American Mass Market*. New York: Pantheon Books, 1989.

Strom, Sharon Hartman. *Beyond the Typewriter: Gender, Class and the Origins of Modern American Office Work, 1900–1930*. Urbana, IL: University of Illinois Press, 1994.

Tallman, George and Bruce Blomstrom. "Self-Service in Soft Goods." *Journal of Marketing* 25 (1961): 133–43.

Unger, J. "Self-Service Shops and the Law of Contract." *The Modern Law Review* 16 (1953): 369–70.

Wertenbaker, Charles. "Patterson's Marvelous Money Box," *Saturday Evening Post*, October 3,1953.

Zimmerman, M. M. "The Supermarket and the Changing Retail Structure." *Journal of Marketing* 5 (1941): 402–9.

Zimmerman, M. M. *The Supermarket: A Revolution in Distribution*. New York: McGraw-Hill, 1955.

Zukin, Sharon. *Point of Purchase: How Shopping Changed American Culture*. New York: Routledge, 2004.

# 2 Phantom of the Operator[1]
## Rotary Dialing and the Automation of Everyday Life

When the telephone arrived in Kansas City, Missouri, during the late 1880s, service was erratic. Sound quality was poor, disconnections and wrong numbers were frequent, and inexperienced operators could be as unreliable as the new technology. Some early subscribers doubted operators' integrity as well as their competence. When an undertaker named Almon B. Strowger noticed a dip in his business, he began to suspect that local operators were in cahoots with one of his rivals. He became convinced of conspiracy once the family of a deceased friend told him they had been connected to a competitor after asking for Strowger by name. The undertaker was already a disgruntled subscriber, because he felt operators were overly prone to mistakes; furthermore, when operators got flustered or frustrated, which occurred often during the telephone's early days, they could be inconsiderate or even rude. Now suspecting collusion as well, Strowger took matters into his own hands. He began working on a fully automatic telephone exchange system, which would eliminate the risks and frustrations of relying on operators to conduct his business.

The telephone itself was barely a decade old when Almon Strowger began experimenting with automated connections. The fledging Bell Telephone Company (named for its founder, Alexander Graham Bell) was still struggling to convince Americans to use the new communication technology. Initially, the telephone was saddled with the nickname "Bell's Electric Toy."[2] In part to overcome this stereotype, the company directed the bulk of its early promotion toward businesses reliant on the telegraph. The telephone's advantages were immediately apparent; most notably, vocal dialogue vastly increased the ease and possible range of expression. Operators, however, were also freed from the restrictions of Morse code, now able – indeed required – to interact with customers while placing their calls. Virtually all telegraph operators in the U.S. had been men, and a "quasi-craft model of work" developed around the technical, manly world of Morse code, which largely insulated them from suspicion of occupational hazards like eavesdropping or intentionally misdirecting calls.[3] No such model developed to describe the labor of telephone operators, and for some early subscribers like Strowger, the new female presence on the line compromised the potential of the new technology itself.

Several telephone companies were already experimenting with automatic systems when Strowger decided it was time to do away with operators. The American Bell Telephone Company was not yet the AT&T monopoly it would become, but it already dominated the telephone industry. Its engineers and patent lawyers kept watch over all innovations and developments in automated connections, although the Bell Company steadfastly refused to introduce any elements of automation into its own service. As the industry leader for a new technology, a primary task of the ascendant Bell Company was convincing people to use the phone, still largely considered a novelty in the 1880s. In 1887, for example, at an annual conference about the switchboard technology used by operators to connect calls, a Bell executive explained his resistance to automation in plain terms: "it makes it [telephony] hard, and I should say that we should make the operation of using the telephone as easy as we can for the party who is to use it."[4] "User transparency" was a house principle for the Bell Company during the telephone industry's formative era. It was the unproven (and unreliable) telephone that Bell wanted to make "transparent" for its customers, so the company offered interactions with operators in place of callers having to struggle with the new technology.

Bell's founders opted for operators in the first place, rather than automation, as a strategy to attract subscribers. Operators became the means by which Bell distinguished their brand of telephone service from upstarts and independent competitors. The nascent monopoly's brass poured its superior resources into cultivating a personalized relationship between their operators and their customers. Many subscribers grew attached to their "hello girls," as "Ma Bell" paternalistically referred to "her" operators in advertisements and promotional literature. When people began dialing their own phones, they also said goodbye to their "hello girls." The humble rotary dial is already outmoded, first replaced by touch-tone keypads (the subject of the following chapter) and more recently touch screens, yet dialing remains an historically transformative consumer labor technology. And no less than self-payment or telephony itself, dialing constituted a new assemblage in its own right, replete with human as well as mechanical technology.

Several accounts of telephone automation begin with Almon Strowger, and I rehearse the origin story to demonstrate that a history of dialing in the U.S. cannot help but feature the Bell Telephone Company and the women employed by Bell as operators. In the first place, Strowger was driven to develop an automatic dialing system after becoming intolerant of operators. Bell would develop into the AT&T monopoly and control American telephony for the lion's share of the twentieth century, but the first famous attempt to automate telephone connections in the U.S. stemmed from one customer's dissatisfaction with Bell's operator service. Meanwhile, Bell's longstanding aversion to automation was rooted in a strategy of "user transparency" for the telephone, which they employed operators to maintain. Strowger's automated system became legendary because of the personality of its namesake; technologically, the Strowger design was poorly devised. It required much more wiring and battery power than other early automatic versions, which meant significantly higher installation and maintenance costs. Strowger's system also offered a cumbersome form of interface that required users to manipulate a series of buttons to place a call. The clunky Strowger system highlights why, when Bell finally automated its own system, the company adopted the rotary dial – it was the easiest interface for callers to use.[5]

Despite its flaws, Strowger patented his automatic system and then wrote the Bell Company to inquire about selling his patent. Strowger's prototype was reviewed by Bell's engineers, along with all new developments and innovations, and Bell's patent attorney reported to the company president that "[t]here is nothing practical or meritorious about the particular form of apparatus described in this patent. ... The [Strowger] patent would be of no use to this Company or any of its affiliates."[6] Given the Bell Company's guiding principle of "user transparency," its executives were probably more averse to the complicated push-button system than the added start-up costs. Undeterred (and, according to the legend, probably spurred) by the fact that Bell never dignified his solicitation with a reply, Strowger recruited investors and joined forces with other inventors to fix the problems of his original design. Together, they formed the Strowger Automatic Telephone Exchange, which incorporated on October 30, 1891, and opened its first commercial exchange in LaPorte, Indiana a year later. The Strowger exchange grew over the next twenty years to become "the largest and most successful automatic telephone company and telephone equipment manufacturer in the United States."[7] Like nearly all independent companies offering automatic service, Strowger went out of business shortly after World War One, once Bell finally decided to automate its own system.

When Bell began automating its first local exchanges, automation was understood to be spilling out of factories into American homes and offices. During the first two decades of the twentieth century, "[m]anufacturers realized that machines could take over the tasks of filling cereal boxes, rolling cigarettes, or even twisting pretzels. In automaking and elsewhere, technology seemed to substitute for people's skill and experience."[8] The dial was similar to these examples of automation and also different. On the one hand, dialing fit the pattern of new workplace machinery replacing paid labor; but, on the other hand, operators were employed to provide a service rather than to manufacture a product. Dialing was not a case of automation whereby machines were simply or entirely "tak[ing] over the tasks" of employees. The automation of telephone connections eliminated some tasks performed by operators, while other tasks were reassigned to callers. Service is by definition relational, and the production of goods can be automated to an extent that the provision of service never can. The automation of telephone connections eliminated some tasks performed by paid employees, namely operators, but other tasks were also reassigned to callers.

When Bell's subscribers began dialing, it also meant saying goodbye to the company's "hello girls" (as operators were condescendingly but affectionately known) and assuming responsibility for their primary function.

Amy Sue Bix's history of public policy and debate about "technological unemployment" elaborates the broader national context within which telephone automation occurred. Although Bix focuses on blue-collar manufacturing sectors of the economy, operators and the dial are the first example she mentions of paid labor being replaced by new machinery. Bix begins her history in 1929, and she describes how the national attitude toward automation flipped with the onset of the Great Depression. "In the 1920s, mechanizing production had seemed to guarantee prosperity; during the 1930s, people feared that changing workplace technology might become America's social and economic downfall."[9] Bix explains that the technological unemployment "became vital to Depression-era thought precisely because concern stretched across occupational lines to become a nationwide controversy."[10] Bix highlights the rotary dial because it was the first case of labor automation to generate widespread public concern for white-collar (or, more precisely, pink-collar) workers, alongside blue-collar workers. Furthermore, beyond Bix's debate, dialing was an unprecedented case of technological displacement because it engendered concern among the dialing public about consumption as well as jobs, specifically the consumption of telephone service.

Accounts of telephone automation tend to focus on operators; in this chapter, I approach dialing from the other end of the line. Venus Green's history of "gender, labor and technology in the Bell System" provides the authoritative account of automation's impact on operators, and Steven Norwood's labor history of operators describes their reactions and organized resistance to automation.[11] In this chapter I narrate a dialer's history of telephone automation, by focusing on callers' experiences of the same processes. In her history of department store clerks, Susan Porter Benson describes the customer as "usually a phantom to the production worker [yet] a continuing presence in the life of the saleswoman."[12] Here I reverse Benson's insight to analyze how callers began using the telephone in the absence of operators. Building on the history of self-service shopping presented in Chapter 1, Chapter 2, now begins to address the most fundamental question in this book: how did the naturalization of consumer labor in the U.S. inform subsequent attitudes and expectations about interacting with everyday technology?

By 1930, nearly one-third of all telephones in the U.S. featured rotary dials. That spring, when dial phones were installed in the U.S. Capitol building, resolutions were introduced on the floor of the Senate and the House to ban the dial in all Congressional offices. On the cusp of the Great Depression, the number of women employed as operators was plummeting, but the protest on Capitol Hill was not organized in support of working women. Carter Glass (D-Virginia) sponsored the Senate resolution on his own behalf, to avoid being "transformed into a telephone operator myself without compensation." Today the dial hardly sounds like cause for alarm or something to resist, but the resolutions "launch[ed] a citywide debate among elected officials, businessmen of all kinds, and local residents" about the merits of the dial and who should be conscripted to use one. An amendment was finally added to the resolutions "allow[ing] the dial telephones to stay except in the offices of Senators who specifically wish their removal." Dialing became second-nature, of course, but this forgotten controversy reminds us that people often need to be convinced to take on even minor new tasks, especially, as Senator Glass made clear, when replacing other people who had been on hand to perform that task for you.[13]

In this chapter's first two sections I detail Bell's preparation of its subscribers for rotary dialing, drawing on histories of operator labor informed by the monopoly's archives. To capture a sense of dialing's novelty I also analyze instructional films Bell used to publicize dial telephones and the new task of dialing. These films demonstrate how much time and energy Bell devoted to its mission of convincing subscribers that dialing was an improvement, rather than a reduction, in their service. I also searched for moments of organized resistance to systematic changes in callers' experience of telephony. Resistance since the dial has been scarce, so I describe more fleeting and individual moments of protest as well. In the third and final section, I explore the social implications of callers' personal attachments to operators by comparing them with later attachments callers made to their local exchange names, such as "BUtterfield-8," and later still to area codes. The retirement of exchange names in favor of "all-digit dialing" met with relatively intense protest, one of the only sparks in the history of subscriber resistance to find any oxygen. Both collectively and individually, callers' emotional connections to their telephone numbers, it turns out, bear traces of the class and status dynamics that had animated their social relations with operators. During the dial's formative era telephone subscribers struggled less with the new task of dialing than they did with the loss of these gendered, raced and classed relationships. The history of dialing highlights how, even (or perhaps especially) in the absence of employees, consumer labor assemblages still involve articulations between the servers and the served.

## The Slow Road to Automation

The telephone was already an established domestic technology when Bell decided to automate its local connections. In 1919, when Western Electric (Bell's manufacturing subsidiary) rolled out its first rotary model, more Americans had residential telephones than had automobiles, and there were more than twice as many telephones as radios in homes across the U.S.[14] When Americans began using shopping carts and automobiles to haul their own groceries home, many of those homes were outfitted with AT&T telephones, and subscribers were already beginning to use rotary dial telephones to place local calls. The history of self-service, described in the previous chapter, reaches back to the Great Depression, and by the end of the 1930s, supermarkets were the dominant form of retail grocery distribution in the U.S, and self-service was beginning to spread across other retail sectors. Dial telephones began appearing in American homes and offices a full decade before self-service won approval as a retail distribution strategy.

People did not begin dialing their own phones until after World War One, but automated connections are nearly as old as manual connections. The first automatic system was patented in 1879, less than a year after the first successful manual exchange had opened for business (and a decade before Almon Strowger received his first patent). Before the turn of the twentieth century, scores of automatic telephone systems were invented and developed. Bell reviewed the patents for every new system, and even purchased a few that were considered to be potentially threatening to the company's emerging brand of operator service, marketed with the (patented) slogan, "the voice with a smile." Bell's executives remained convinced that manual switching and operator service paved the surest path to dominance of U.S. telephony, but as early as 1886 the company assigned several engineers and technicians to begin developing an automatic system. From its origins, Bell seemed resigned to the fact that automatic telephone connections were inevitable; however, the company would be in business for over forty years before it began asking its customers to dial their own phones.

The first telephone companies competed with the telegraph, which was the dominant (indeed, the only) form of fast, interpersonal telecommunication available. In order to prove its superiority over the telegraph, "the telephone industry realized that it would have to expand the functions of a telephone exchange beyond a simple connection."[15] Telephone companies targeted businesses reliant on telegraphy, and in commercial markets "specialized attention quickly developed into a profitable means of attracting new customers." Telephone companies were willing to let their business customers determine what they needed, and operators were trained and instructed to cater to subscribers' individual desires and demands. Operators not only connected calls; they also helped track down clients or colleagues, delivered messages and performed other quasi-secretarial functions. Not surprisingly, businessmen were also the first telephone customers to install telephones in their homes, where they (and their wives) expected the same pliant personal service received at the office. At the office, this service took the form of an additional secretary, while at home the model was the all-purpose domestic servant. For home subscribers the phone – and its attendant operators – emerged as a status symbol as well as a handy new tool. "Bourgeois subscribers, accustomed to having servants, bought telephones based on the expectation that operators would serve [them]." Telephone users at home and at work alike were trained to expect that the new technology, unlike the telegraph, brought with it an expansive form of servitude, and in large part these expectations account for the long, slow road to automated telephony in the U.S.

In its earliest years, before it grew to dominate the industry, the Bell Telephone Company focused on winning large, urban markets, and its managers and executives remained skeptical that automatic systems could be technologically viable in cities. Automatic exchanges suffered from a diseconomy of scale, meaning that they became more expensive to use as the number of switchboards and the amount of traffic moving through them increased. Many of these higher costs were labor costs specifically: "in exchange for the elimination of operators who earned small salaries, the first automatics offered higher installation, production, and maintenance costs of the wages paid to skilled craftsmen, unreliable equipment, and limited types of services."[16] It would be thirty years before Bell would finally decide that the initial costs were worth the long-term savings offered by automatic systems. Bell was willing to install automatic systems in rural areas much earlier than in urban markets, as the lower traffic rates in small towns could not even pay for the salary of one operator working full-time, let alone the staff required to maintain twenty-four hour service. Even in these rural markets, however, Bell considered an automatic system to be a temporary solution, and they were replaced with manual systems as soon as the exchange grew enough to support a full-service system staffed by operators. Historians of technology tend to assume that Bell began promoting dial connections as soon as "the switchboard problem" was solved,[17] but labor historians focused on operators (rather than engineers) have uncovered a much more complicated process, whereby automatic connections had to become socially and culturally acceptable as well as technologically possible and commercially viable before Bell would begin installing rotary dial telephones in homes and offices across the U.S. It was not simply a question of Bell's engineers making automation possible on a wide scale; it was also a question of Bell's managers and marketers persuading its consumers to accept dialing their own phones in place of operator service.

Independents such as Strowger tried to market themselves in distinction to Bell by highlighting the unreliability of operators. For example, in addition to promoting such advantages as "immediate connection," "continuous service," and "lower rental" rates, one early advertisement for the Automatic Telephone and Electric Company of Illinois also promised to eliminate from telephone service the "frequent tedious delays occasioned in indifference of operators or [their] inability to handle business as promptly as desired." Furthermore, echoing Strowger's original concerns, the ad guaranteed "[a]bsolute secrecy of conversation." Many early independent companies offering automated service "made similar promises, and they all emphasized the elimination of operators" when describing the advantages offered by their automatic systems, chiefly lower costs and better service.[18] In contrast, Bell's early publicity routinely touted manual systems as experientially superior because of the advantages offered by operator service. (Internal deliberations about automation, meanwhile, focused on costs, for instance efforts to pinpoint the rate of growth required to make it worth absorbing the initial sunk costs of setting up an automatic system in exchange for future profits stemming from reduced operator payrolls.) Some publicity made simple (if absurd) arguments, such as, "[An] automatic system cannot be operated by the subscriber in the dark." Bell affiliates also touted the extraordinary abilities of operators and their capacity to assist callers during emergencies. In a legal brief challenging an automatic competitor, the Southern New England Telephone Company argued that the "the manual operator has been able to summon assistance at all hours, day and night, and to bring to the aid of the subscriber, in such cases, the help of neighbors, of the police or of the fire department or of all of them together." Often the gendered and class-based notions of servitude, so central to Bell's promotion of operator service, would be accompanied by a corresponding notion of female helplessness. When Bell started marketing to home as well as office users, the company had it both ways in some ads, claiming that residential telephone users (overwhelmingly women) would be unable in an emergency or crisis to manipulate a telephone correctly, and instead should rely on operators (also overwhelmingly women) to calmly and reliably perform these functions for them.[19]

Figure 2.1 "Even in today's rush and hurry..."[20]

Between the turn of the twentieth century and the outbreak of World War I, Bell gradually improved its technology and developed semi-automatic systems for connecting calls. The subscriber's experience of telephony began to change over this period, as Bell introduced features such as automatic ringing, busy signals and, most significantly in terms of paving the way for the operator's extinction, the dial tone. By 1917, Bell has installed fully automatic switching systems in all but their largest markets – New York City, Chicago, Philadelphia, and Boston. However, Bell maintained full operator service in all markets, even as they began upgrading to automatic switching technology. By bringing in automated equipment behind the scenes, if you will, Bell was able to "bring savings in cost ... without requiring work by the subscriber."[21] In other words, Bell adopted its first automated technology to make an operator's labor more efficient, rather than to do away with it (yet). The customer's experience of Bell's service still very much centered on the operator, even while the technology was being developed and installed that would make it possible to automate telephone connections.

Until World War I, Bell's switching equipment was no better than many of its independent competitors, and the company relied on operators to make up for malfunctions and inconsistent service. As Venus Green writes, "[f]aced with competition and severe equipment problems, Bell executives quickly transformed auxiliary services into 'personal service' as a means to capture and dominate the industry." Bell's first famous advertising slogan, "the voice with a smile," trumpets operators as what Bell was offering, rather than the telephone itself. Ironically, Bell became a victim of its own success, and in order to persuade its subscribers to begin connecting their own calls, the company would have to invest tremendous time and resources into undoing the service relationship they had so carefully established. Green developed the concept of "distancing" to describe how Bell paved the way for the dial by gradually withdrawing elements of personalized operator service. The "Bell System progressively disassembled its long-established cultural links with subscribers by eliminating much of the customer contact and many of the free services operators formally provided." In "I Was Your Old Hello Girl," a nostalgia piece published in *The Saturday Evening Post* in 1930, former operator Katherine Schmitt wistfully recalled how, during the formative era of operator service, the range and unpredictability of callers' questions, beyond straightforward requests to connect calls, had made her job "a combination of personal service bureau and general guide to the city." Then, in the decades around the turn of the twentieth century, operators' jobs "changed dramatically as management applied the principles of Taylorism to switchboard work." However, it was not a clear, inverse relationship between standardization and personal service; it is more accurate to say that Bell first used scientific management to standardize its personal service, and then gradually and systematically removed features of that service in anticipation of automation. Operators lost control over their own interactions with callers long before they stopping having them.[22]

By the turn of the twentieth century, "the degree of supervision exercised over the telephone operator was unparalleled in any other occupation."[23] In a nutshell, efficiency replaced customer satisfaction as the hallmark of Bell's service. Interactions between operators and subscribers became "formal and impersonal," as Bell imposed rigid protocols and limited phraseology to restrict operators' ability to chat with or assist callers in any depth. The primary locus of control for Bell managers became what operators were permitted to say. Earlier recruitment, training and supervision had focused on how operators spoke: a cheery tone and crisp, clear pronunciation were paramount. (These job requirements helped justify Bell's notoriously exclusionary hiring practices, under the pretense that African-Americans, Jewish women, and other minorities were prevented by accents and dialects from embodying the proper "voice with a smile.") Standardization shifted supervisory scrutiny from how operators spoke to what they could say. By 1912, operating phraseology was uniform throughout the Bell System, and operators began answering every call with the question, "Number, please?" This routinization marked perhaps the most significant step toward automation, as operators, in essence, became human dial tones.

World War I was the tipping point in Bell's slow journey to full automation. Labor and capital from the telephone industry were diverted to the war effort, and "[c]onsequently, service deteriorated everywhere," regardless of how far along a particular exchange may have been toward fully automated connections. In 1918, the federal government seized control of telephone service in the U.S. The Post Office Department was assigned control over telephony, and while the era of government control was brief, lasting less than a year, federal oversight constituted "a reign of terror" for operators and their unions.[24] Labor historians like Norwood and Green have documented operators' organized resistance to government control (and union busting), and several operator unions participated in the "unprecedented wave of mass strikes that erupted throughout the country in the year after the armistice."[25] The Bell monopoly was largely consolidated in the aftermath of the war, and this militancy among operators further motivated Bell's managers to automate their local exchanges.

The technology and economies of scale had been in place for decades, but it was not until Bell had vanquished its competition (with help from the federal government) that they accelerated their "progressive dismantling of personalized service."[27] Bell took advantage of the conditions (and expectations) created by the war to withdraw auxiliary services from its subscribers. For example, operators would no longer provide the proper time of day free of charge, let alone weather forecasts or sports scores on demand, as had been earlier encouraged. Now, operators were trained to answer any such questions by replying, "I'm sorry, we do not have that information." Operators also stopped monitoring calls for busy signals and no answers. Heretofore, operators had passed such information onto

**HER BIGGEST JOB**

**IS WAR**

There has never been a time when the work of the telephone operator has been so important as right now.

For there are more Long Distance calls than ever before. More are in a hurry, particularly the urgent calls of war.

Calm in emergencies, capable and courteous, the telephone operators are earning a nation's thanks for a job well done.

**BELL TELEPHONE SYSTEM**

*When you're calling over war-busy lines, the Long Distance operator may ask you to "please limit your call to 5 minutes." That's to help more calls get through during rush periods.*

*Figure 2.2* "... the urgent calls of war."[26]

the caller, but now it became the callers' responsibility to listen for busy signals, or to determine on their own how many rings meant that no one was going to answer. As these examples demonstrate, the advent of automated telephony entailed training callers in a multi-faceted process, and as second nature as dialing may have become, automated connections involved a series of new, interrelated tasks and responsibilities for callers. Despite the experiential complexity of dialing, the diminished expectations of service during the war finally convinced Bell that its customers were ready for automated telephony. Callers had already essentially said goodbye to their hello girls, and they had gotten used to using a telephone without the service relationship that Bell had cultivated between operators and callers. Amy Sue Bix points out that Bell's promotion of its service after the war (and well into the 1930s) continued to highlight operators and the full service model; yet mechanically and experientially Bell had begun well before the war to "distance" their callers from the operators they had come to rely on.[28]

An advertisement featuring an operator, just after World War I, when AT&T was beginning its turn to machine methods of switching. Courtesy of AT&T Company Archives, Warren, New Jersey.

*Figure 2.3* Operators still loomed large in Bell's marketing after the war.

Throughout the 1920s, the number of women employed as operators actually increased, even as dial phones were being introduced. Conversion to dial during the 1920s proceeded at "a slow to moderate pace," but the increase in dial phones over this decade was substantial.[29] In 1920, only two percent of telephones in the U.S. were dial; by 1929, over one-quarter of telephones were dial. The telephone itself was becoming an indispensable feature of home and office life in the U.S., and "rapid telephone expansion in the 1920s created a higher demand for operators than dial phones eliminated."[30] These macroeconomic factors, and a reliance of operator employment rolls, help explain how in 1930 the Department of Labor could conclude that Bell was doing "a model job in smoothing the transition from manual to dial" telephone connections.[31] Bell won this approval for its approach to automation in no small part because they had organized the switch to the dial to be gradual, rather than all at once. Bell invested vast amounts of money and manpower into the conversion to dial, over decades, and the slow implementation was claimed to help soften the landing of laid-off operators. The gradual approach was also meant to ease the telephoning public's transition to the dial.

In 1933, the D.O.L issued a "Bulletin of the Women's Bureau" focused on "the change from manual to dial operation in the telephone industry." The report is expectedly focused on the problem of laid-off operators, and it pays close attention to Bell's efforts to help relocate and support expendable employees. However, the bulletin also acknowledges the dual challenge Bell faced of not only justifying operator layoffs, but also convincing its customers to accept the new tasks of automation. The bulletin assumed that the telephoning public was not only aware of the plight of operators, but also filled with inflated fears about how complete a change the dial would bring. A technological upgrade like dial connections was "not one to be easily understood by the telephoning public," and to "the ordinary telephone user it seems that everything now is mechanical, whereas formally it was human," despite the fact actually there were still tens of thousands of operators employed to perform a range of functions including the connection of all long-distance calls. During the Great Depression sensitivity to automation was heightened because of fears about unemployment in most sectors of the economy. But as the Labor Department report illustrates, with its description of the telephoning public's concern that "everything now is mechanical," callers also regretted parting with operators because of the everyday (and often intimate) interactions they were used to having with them. Bell was not only reorganizing the standard practice of telephony; they were also dismantling the brand of operator service – the patented "voice with a smile" – that the company itself had spent decades cultivating in its employees and marketing to its customers. The Women's Bureau bulletin about telephone automation acknowledges the dial would mean tremendous changes for the telephoning public, as well for as the operators who had been employed to serve them. However, the implications of automated telephony listed in the bulletin do not extend beyond the switchboard. Hence, while the dial represented familiar challenges for operators, the only "results to the consumer" mentioned describe a series of advantages. The dial system will be faster, more efficient and consistent, and "just as rapid during the busier hours of the day." Furthermore, compared to operators, the new dials will be "more adaptable to improvements and developments" and better equipped to "keep pace with the rapidly increasing requirements of the telephone service." The bulletin goes to great lengths to reassure its readers that operators have been well provided for during the transition to dial connections, yet at least it presumes that these concerns were reasonable. Callers' concerns about what dialing would mean for themselves, on the other hand, are presented as misguided.[32]

It is no surprise for a report from the Labor Department to maintain a framework for understanding automation as a burden that falls on employees. From a perspective oriented toward labor, automation entails processes by which machines replace paid employees. While this may be true for cases of automation in manufacturing sectors, by and large, the automation of service occupations reveals a more complex situation, in which some of the paid labor is usually transferred to the customers themselves. When customer service jobs are automated, the work seldom if ever disappears, and the shift from operators to rotary dials was an early example of customers taking on portions of what had been a service job, until the position was "automated" away. Not unlike the bulletin from the Women's Bureau about "the change from manual to dial operation in the telephone industry," the labor histories I draw on in this chapter focus on how the change impacted operators. In this section I have mined these sources anew in order to describe how telephone companies managed automation for callers as well as operators. In the next section, I turn to the promotional and instructional films Bell produced to introduce the dial to its subscribers.

## Lights, Camera ... Dial!

After World War I, enough foundation had been laid for the Bell Company to begin rolling out dial telephones. The first fully automated system was installed in Omaha, Nebraska in 1921. Long-distance calls placed in Omaha would still require manual operation for years, and nationally the gradual shift from operator service to full automation for all calls continued well after the introduction of the dial. Demand for telephone service rose steadily during the 1920s, and in some new markets local calls were connected automatically from the get-go, while other, older markets would have to be converted. The last manual long-distance board, in New York City, by far Bell's biggest market, was not fully converted to dial until 1960, but all local calls in New York were connected automatically by callers by the time the U.S. entered World War II. Numerically, the 1930s were the decade during which automatic telephone service reached critical mass in the U.S. In 1930, home telephones were roughly as common as cars and radios, and less than one-third of telephones in the U.S. were dial; by the end of the decade, the Depression had caused the number of telephones across the country to dip slightly (and temporarily), but by 1940 over fifty-five percent of phones in the U.S. were dial. Over the same decade that most Americans began selecting and transporting their own groceries, subscribers to telephone service also began using the rotary dial to place local calls.[33]

During the 1920s and 1930s, Bell began promoting dial telephones directly to its customers, one market at a time. First, the local Bell subsidiary mailed a series of notices to all of its customers, culminating in the announcement of the "cutover," the exact time and date when their local exchange would convert to a dial system. Western Electric, AT&T's manufacturing subsidiary responsible for building the new dial phones, also began producing instructional films to help introduce the dial to the telephoning public. Like the individual notices, these films were intended to reach subscribers to each of Bell's local exchanges. For example, in 1927, Pacific Telegraph & Telephone screened the seven-minute silent film "How to Use the Dial Phone" for subscribers in Fresno, California. The film offers a template of Bell's promotional tactics for the dial in markets nationwide, but the target audience is also remarkably specific: the first title cards reiterate "Midnight, May 28, 1927," as the moment when Bell customers in Fresno would no longer place their local calls with an operator. From that day forward, local calls would be connected directly by the caller. The significance in the film of establishing the exact date of the "cutover" highlights the challenge Bell faced in introducing such a drastic change into the basic experience of telephony. Despite the decades of planning and preparation, and no matter how much "distancing" Bell had achieved between operators and their customers, the actual shift from manual to dial connections for local calls took place in one fell swoop. Literally in an instant, a familiar service became a new task, or actually a series of tasks, as films like "How to Use the Dial Phone" demonstrated. Customers' experience of the "cutover" helps explain why Bell had planned the lead-up to this moment over decades.

NEW YORK TELEPHONE COMPANY

204 SECOND AVENUE, NEW YORK

EXCHANGE 5-4500

November 21, 1941

Dear Subscriber:

        To extend the advantages of direct dialing of telephone calls, new equipment has been installed in your central office so that after 8 A.M., Sunday morning, November 30, you can dial directly calls to all points in New York City.

        As you know, it has been necessary to dial the operator when calling the more distant points in the city. But on and after November 30, this will no longer be necessary and all such calls should be dialed.

        With this improved plan, it will be still easier to make your calls. There will be no change in charges and your bill will be sent you in the same form as at present.

        The names of all the central offices and the charge to each are shown on the enclosed card. If you would like to have additional copies of this card or any further information, just dial "811" and ask for the representative handling your account.

NEW YORK TELEPHONE COMPANY

Remember the time and date - Sunday morning at 8 - November 30, 1941.

*Figure 2.4* Announcing the "cutover."

The suddenness of the "cutover" also helps explain why the Women's Bureau bulletin (issued six years after the Fresno film) would complain that the telephone public was overestimating how complete a change the dial actually meant for telephone service. The film, like the bulletin, emphasizes that operators will still be on hand to place long-distance calls, troubleshoot, and to provide a new function, "directory assistance," which callers could now ask for by dialing "8." The film reassures viewers that an operator could also still be accessed by dialing "0" for assistance. Given Bell's consistent concerns about minimizing work for the caller and making calls as simple as possible, it is notable that they provide two different numbers for access to operators. Automation not only brought about an elimination (technically, a reduction) in operator service, it also meant that an operator's job was fragmented into a set of discrete and finite functions, in stark opposition to the all-purpose service Bell had originally trained operators to provide.

"Directory assistance" marked the introduction of a new service provided by operators, the film stresses that callers should only dial "8" as a last resort. Bell's "Blue Book," preceded the iconic Yellow Pages as a new technology of consumer labor within the assemblage of dialing. Looking up a telephone number was another new task concomitant to dialing itself, a caller's first step toward the successful completion of an automatic local connection. The film "How to Use a Dial Phone" represents this initial step in great detail, with an animated lady getting out her blue book, followed by a live-action counterpart looking up a number. Title cards narrate each step. Once the number is retrieved, the second step is to "remove the receiver carefully so as not to jiggle the switch hook." Next, the caller must "listen for the dial tone," and a separate title card explains that the dial tone is a "steady humming sound indicating that the line is ready for you to use." Finally, several cards interspersed with parallel movements by the two figures comprise the final step: actually dialing the phone, one digit at a time. Once the call is complete, a card triumphantly reports, "You have dialed 3-6623. Note: the dash (-) is not dialed." Finally, a reminder to replace the receiver as carefully as you picked it up, again "so as not to jiggle the switch hook."

The painstaking, step-by-step presentation is tedious and surreal. It also dramatizes how entirely new dialing was for Fresno residents in 1927. The film reviews several of the potential pitfalls facing dialers on the way to a successful call, with a series of titles warning, "You will get the wrong number IF--." Operators do not appear in the film until immediately after the list of possible human malfunctions that can impede an automated connection. Their presence is meant to be reassuring, but it is also clear that their role has diminished. Callers are expected to rely on operators as merely one tool among several offered to them by the phone company. The film emphasizes, for instance, that "directory assistance" from operators would now be secondary to the "Blue Book."

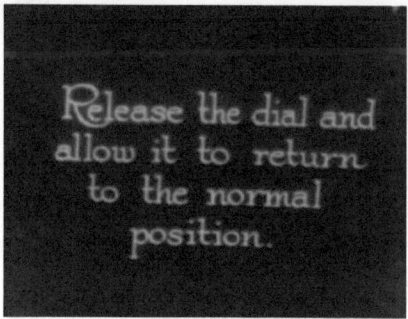

*Figures 2.5–2.7 How to Use the Dial Phone.*[34]

The new assemblage of dialing reassigned the work of telephone connections from operators to callers. Then the protocols of telephone use in the U.S. remained largely unchanged until the proliferation of cell phones more than half a century later. The many technologies required for self-service shopping included shopping carts, standardized cash registers, and new shelving for mass display. More expansively, brand name advertising familiarized shoppers with goods before they entered the store, and shoppers used their own automobiles to transport groceries home. Similarly, while automated telephony revolved around the rotary dial, there were several new technologies introduced along with the dial (some well before it) that made dialing manageable for callers and for local telephone exchanges. Automatic ringing, busy signals, and, most significantly, the dial tone absorbed aspects of operator labor before callers actually began using rotary dials to place their own calls. The dial tone, like the rotary dial itself, marks a significant moment in the history of automated consumption. The dial tone domesticated the work whistle, as consumers were trained to realize that an aural prompt meant it was time for them to begin working, more specifically that it was time to dial their phone.

Automated connections meant that callers assumed responsibility for learning each other's telephone numbers. Telephone numbers preceded the dial, and in a sense connections were automated at the point when operators'

Here's an Attractive
Booklet for Your
Telephone Numbers

We've just printed a new booklet
for listing your personal telephone
numbers.

You'll find you save time on Long
Distance calls when you give the
operator the number you want.
This booklet helps you keep the numbers handy.

*SAVE TIME ... CALL BY NUMBER*
So use the booklet to write down the
numbers you already know. If there's
a new number you don't have — or an old
one you've forgotten — be sure to add it to
the list when the operator gives it to you.

There's a copy for you at the near-
est Bell telephone office.

**BELL TELEPHONE SYSTEM**

*Figure 2.8* Before the Yellow Pages, Bell's Blue Book.[35]

interactions with callers were standardized into the routine phrase, "number, please?" Likewise, arguably the most significant imposition on dialers
was not the physical manipulation of the dial itself, but the responsibility of
storing or remembering each other's numbers. Extensive testing found that
the average customer could remember no more than six digits at a time,
and the larger urban exchanges required seven digits per telephone number
in order to facilitate switching for all numbers in the city. AT&T engineer
W. G. Blauvelt came up with the idea of combining the three letters of the
exchange name, which callers were already familiar with, with four additional numbers that could much more easily remembered. The result was the
system of exchange names and numbers that comprised telephone "numbers" for the first few decades of automated dialing. As exchanges grew in
size, a fifth number was added, a single digit combined with the exchange

name, so now that telephone exchanges would be identified, for example, as "BUtterfield-8." As the next section of this chapter discusses in detail, callers would not only accept these exchange names, but many would also embrace them, and some would even become as attached to them as telephone subscribers of an earlier era became to operators.

By the onset of World War II, well over half of telephones in the U.S. were dial, but the manufacturing subsidiary Western Electric continued to produce promotional films designed to help Bell's customers accept the dial and learn how to use it. Earlier films, like the Fresno example, were produced for specific local exchanges, as Bell assumed that each and every telephone user could use some hand-holding during the first waves of cutovers. Later films continued to target particular markets, but now, rather than focus on one exchange at a time, Bell isolated particular demographics that were relative latecomers to the dial. One such group were the elderly, routinely assumed (then as now) to be more "set in their ways," resistant to change generally and skeptical of new technology specifically. Bell executives also worried that rural populations would be particularly resistant to the dial. This concern was somewhat ironic, given that rural markets were often first introduced to automatic systems before they grew large enough to support operators' salaries, as mentioned earlier. However, the smaller scale operations in rural areas meant that callers and operators were more likely to get to know one another personally. (They were also more likely to have known each other personally before they began interacting as callers and operators.) In rural areas, callers' interactions with operators often grew into more intimate and egalitarian relationships than in urban markets, and consequently Bell was concerned that these markets would be particularly resistant to the dial.[36]

A promotional film titled "Dial Comes to Town" targeted rural and older customers simultaneously. Unlike the straightforward demonstration of "How to Use the Dial Phone," this later film couches its instruction and promotion in a fictional narrative that lasts over twenty minutes. Its dramatic arc follows Grandpa Ollie, who initially protests the dial but triumphantly embraces it at the film's conclusion. The first couple scenes introduce Ollie and his family, who all condescendingly humor his resistance to dialing. At one point, the mother character reassures her young daughter, Jenny, that Grandpa will come around soon enough and embrace the new technology. As they set the table for dinner, the mother reminds Jenny that Grandpa "said the same thing when I got my new washer, remember?" "Oh yes," Jenny giggles in reply, "he wouldn't go near it for weeks! He even likes to work it now." The father, more civic-minded and forward-looking – and less concerned with domestic technology –walks in and sits down at the table beaming about what an improvement the new dial phones and their "fast, up-to-date service" will mean for the town. After dinner, the adults in the family proceed to a town-hall meeting where local Bell managers introduce them to the new dial service. (Jenny wants to come, too, but is assured that there will be an assembly at her school tomorrow covering the same material.) At the meeting, two Bell managers introduce the dial to the townspeople by way of comparison to operators. In a nutshell, these two characters personify the dual-role job description for operators as Bell had developed it. On the one hand, as Mr. Johnson makes clear, operators performed the technical function of connecting calls. However, as Mrs. White elaborates, operators also provided a personal service for Bell's customers that exceeded the mechanical task of connecting their calls.

While explaining the cutover, Mr. Johnson spends several minutes saluting all of the hard working Bell employees who helped bring the dial to town, such as the linemen who rewired the townspeople's homes. He adds operators to the end of his list, but not to applaud their contribution to Bell's service. Instead, he baldly equates the new dial tone with an operator's voice, calling them "the same thing" while insinuating that the only difference between them is that a dial tone never takes a break or needs to sleep. Mr. Johnson concludes his presentation with the following facts: "From [now] on, you will have a dependable, electrical operator, at your service, ready at an instance's notice, one that will be there twenty-four hours a day to handle your calls. Instead of hearing the familiar 'Number, please?' you will hear a dial tone, which says the same thing, electrically." Mr. Johnson's presentation marks a complete reversal from Bell's original marketing of operator service. After initially promoting the operator as a personalized service provider, irreplaceable by machines, now Bell was introducing one flat note of electronic drone by equating it to the job title, and the people, whose assignment had been to smile with their voices.

After decades of speaking to an operator each and every time they picked up a telephone, however, Bell's customers would not be so easily convinced of Mr. Johnson's equation between a dial tone and an operator's voice. Cue Mrs. White, who takes the opposite tack. She begins by reminding the audience that "electrical operators" have not done away with the need for "human operators," and she announces that Bell has more operators on the payroll than ever before. She also lists all the reasons why callers will still require operator service: "you'll still be placing out-of town calls, you'll still want us to find unlisted numbers for you, or you may need help in an emergency." "Dial or no dial," she calmly concludes, callers will still have access to operators. "Don't worry," she reassures her audience, "you'll find operators, just as before, when you need them." In the early days of the dial, as Mrs. White announces, operators still connected long-distance calls as well as providing new services necessary for automation, such as directory assistance. But Bell was severely restricting the criteria of what it meant to "need" an operator. A basic local call no longer qualified.

Mrs. White then walks over to a giant mock-up and proceeds through a painstaking demonstration, nearly ten minutes long, of how to place a telephone call using a rotary dial. She focuses, repeatedly, on each step, from looking numbers up in directories to returning the handset after a call's completion. She continues to humanize the new technology, explaining for instance that the "dial tone is the equipment's way of telling you it's ready to put through your call," reiterating but softening Mr. Johnson's equation between the dial tone and an operator's voice. In essence, her demonstration is "How to Use the Dial Phone" personified in one nurturing and reassuring character, as opposed to animated and live-action exemplars. Like the earlier film, after walking her audience through the steps of placing a call, she then also covers common human malfunctions, such as not releasing the dial properly once your finger has reached the "finger stop," or not letting the dial return to its original position between each digit. "Don't try to hurry it back" is Mrs. White's advice. Finally she demonstrates some potentially unexpected outcomes, such as a busy signal or a missed connection.

*Figure 2.9* Larger than life.[37]

Occasionally the film cuts to Grandpa Ollie sitting in the audience among the townspeople, skeptical at first, but rapt by the end of Mrs. White's presentation. After the town-hall meeting, the film ends with Ollie making a local call, while his smiling family looks on from the other room. He flips through the new directory of local numbers, and then slowly, gingerly at first, he begins to dial. "Yes, that's the right noise. Now let me see … 4 … 1. …" After dialing the entire number and waiting a brief moment, his face lights up. "By gum, Ed, this thing isn't hard to work! No, sir, got my call through first I tried. Real nice and clear, too, isn't it?" A triumphant score swells, the

Bell logo appears, and the message is clear: even the Grandpa Ollies of the world can use a dial phone.

As latecomers to the dial like Grandpa Ollie were finally learning to place their own local calls, the last major step toward fully automated telephony was already taking place. Most callers began placing their own long-distance calls during the 1950s. Automation of long-distance connections required a national system whereby each telephone number would be unique. Until this point, with operators making long-distance connections between distant exchanges, two different telephones in two different exchanges could have the same number, because operators could use listings of local exchange names to keep the all the numbers straight. But if callers were going to dial their own long-distance calls, without recourse to operators and their switchboards, then each telephone number in the country must be unique. The solution was the area code. By 1950 over 70% of all AT&T telephone service was dial, and by 1960 it was over 95%. By the 1950s most of Bell's customers knew how to look up a number and could manipulate the dial in order to place a local call. The automation of long-distance calls would be a much faster and smoother process than local automation had been, in no small part because callers had already undergone so much initial training. With the automation of local calls, Bell's customers had to begin dialing. When callers assumed responsibility for long-distance connections as well, the dial's familiarity meant the heavy lifting of automation had already been done. Nonetheless, Bell still went about automating long-distance calls slowly and methodically. Like local automation, the first stages of automating long-distance calls took place "behind the scenes." Years before callers encountered the dial, operators began using new, automated technology to connect calls, and Bell proceeded in a similar fashion for long-distance. "Operator toll dialing," for instance, was an intermediate stage during the automation of long-distance connections. Under this short-lived system, callers still gave their desired number – now including a three digit area code – to an operator, who then connected the call automatically.[38]

A promotional film from circa 1950, titled "Speeding Speech," celebrates "Operator Toll Dialing," and demonstrates Bell's strategy of convincing callers to accept and embrace a little more dialing by promoting their technological innovations. World War II and the Cold War had replaced the Great Depression as the primary framework for national consciousness generally, and for public perceptions toward technology more specifically. During this period, "AT&T aimed to elicit appreciation for its technical sophistication, to get audiences thinking about the scientists and engineers of Bell Labs rather than about its 'hello girls.'"[39] By the time long-distance connections were automated, it was no longer necessary to convince consumers that "electrical operators" resembled "human operators," as Mr. Johnson and Mrs. White had done in "Dial Comes to Town," or that a dial tone "said the same thing" as an operator asking, "Number, please?" Now, with automation accepted and operators resigned to a secondary role, it was only necessary to equate long-distance dialing to local dialing. In "Speeding Speech," a male narrator calmly and authoritatively reports that "the day draws nearer when long-distance calls will go through just like local calls," and the equation helps minimize any imposition callers might suspect in the addition of area codes or the automation of long distance calls.

Films like "Speeding Speech" mark a turning point whereby Bell began selling the telephone explicitly as a technology, brought to the public by its scientists and engineers rather than as a service provided by its operators. "Speeding Speech" opens with a fictional character, a businessman trying to contact a colleague in Chicago. He dials an operator and gives her the number he's trying to reach, and before he can light a cigarette (with a lighter, of course, not a match) the call has gone through. The bulk of the film's ten minutes is then devoted to a "behind the scenes" look at how the call went through so quickly. Before, multiple local operators were necessary to route a long-distance call, but now only one long-distance operator initiated the call. New equipment takes it from there, communicating automatically to connect calls between distant exchanges. "How does a development like this come about?," the narrator asks before answering that a "broad program of research and development covering all aspects of communication is constantly carried on in the Bell Telephone Laboratories." The narrator continues, explaining that in order for such advances to occur, "more men with greater knowledge and skill are needed, so telephone men are going back to school. And of course operators have to be trained to use the new equipment." Operators require training rather than knowledge, and their use of new technology, like the technology itself, has become the purview of scientists and managers.

The very title of "How to Use the Dial Phone" marks it as an instructional film, while "Dial Comes to Town" combined promotion with instruction, or more accurately threaded promotion into instruction. By the time of "operator toll dialing," instruction was no longer necessary. The narrator of "Speeding Speech" explains all of the new equipment and highlights how much research and development went into an innovation like "operator toll dialing;" however, "for the caller," he simply reports that long distance calls are "as easy as a local call." When long-distance connections were automated, callers already knew how to dial. Technically long-distance dialing did introduce more work for callers – three more digits per long-distance call – but the added dialing requirement did not feel like an imposition, especially in exchange for faster connections, which now took place before callers could even light a cigarette. The point of a film like "Speeding Speech" is for customers to marvel at the new technology, not to be cajoled into using it.

Finally, another promotional film for automated long-distance dialing, "The Nation at Your Fingertips," heralded the first fully automated long-distance exchange in the country, in Englewood, New Jersey. Whereas "Speeding Speech" gave the engineers and scientists center stage, "The Nation at Your Fingertips" is more focused on the machinery that made long-distance dialing possible.[40] "The equipment that makes this service possible is among the most complex that man has ever devised. ... But for the Englewood telephone user, it's as easy as dialing a local call." Dialing three-digit area codes is "the only difference" between a long-distance call and a local call, and the lack of attention paid to area codes in the film, as in "Speeding Speech," suggests that Bell was not worried about callers' dialing the extra digits. Area codes, after all, were what put "the nation at your fingertips." The format for "The Nation at Your Fingertips" closely resembles that of "Speeding Speech," although the film opens with a ninety-second introduction to Englewood, presented as an Anytown, USA, and chosen for that reason as the "proving ground" for "questions such as, will customers like the new service?, and will they find it easy to use?" After introducing viewers to Englewood, the camera settles into a domestic scene reminiscent of "Dial Comes to Town," but in this case the multi-generational family is spread out across the country. We watch an elderly woman open a letter from her daughter to learn that her granddaughter has come down with the measles. Her husband fails to reassure her that the illness is not severe, finally suggesting, "if you're worried, why don't you call her up? You've got her number in that little book of yours. Go ahead, call her. You'll feel better." We then see the grandmother consulting her "little book," a peronsalized version of the "Blue Book" from "How to Use the Dial Phone." By 1950, telephone numbers were being added to address books. The suggestion that a phone call will make you feel better about a distant loved one also reiterates Bell most famous advertising slogan, "Reach Out and Touch Someone," which during long-distance automation replaced "The Voice with a Smile" as the monopoly's marketing catch phrase. In the context of this chapter, it is worth nothing that the slogan is phrased as a command, even though in commercials and advertisements, especially in jingle form, it sounds more like an invitation.

After dialing, we hear the grandmother say, "hello, Sally? This is mother," and the narrator interrupts to ask, "Simple, wasn't it? She just picked up her phone and dialed her daughter in San Francisco, California. In a matter of seconds, she's talking with her." The narrator proceeds with a capsule history of telephone use in the U.S., as a new scene silently dramatizes a call from the 1880s, when it was "an adventure to talk to friends on the other side of town." A familiar historical equation is forged between demand and innovation. Dial telephones, for instance, are highlighted as a technological advance needed to "handle the increased volume of calls" during the 1920s. Innovation also leads to increased employment, the narrator insists, and he reports that the rolls of Bell employees have more than doubled in the thirty years since the introduction of dial phones. He offers no breakdown of Bell employees by job title or sector, and the film continues Bell's tradition, noted already, of using increased demand to mask automation as a source of technological unemployment.

Histories of telephony register no protest or resistance to the introduction of area codes. Area codes did not register to callers as a new burden, and the addition of three more digits to dial was negligible in comparison to the imposition of dialing in the first place. Callers had already learned to dial and had accepted local connections as their own responsibility, so the addition of three extra digits in exchange for automatic long-distance connections was a deal that telephone subscribers appear to have been willing to make. By the time long-distance calls were automated, dialing had been thoroughly absorbed into the protocols of telephony, along with dial tones, busy signals and directories. Like the automation of local calls in the first place, long-distance dialing again replaced portions of operator labor with consumer labor; however, by then, callers had grown accustomed to taking over aspects of an operator's job. The ease with which callers accepted area codes was not a product of arithmetic; in other words, it was not the case that callers more easily accepted three more digits, as opposed to the five required initially to place a local call. Rather, the addition of a little more dialing was negligible because dialing itself had already been accepted.

## What's in a Number?

Bell's subscribers learned each other's telephone numbers before they began dialing them. The implementation of "multiple switchboards" in the 1880s and 1890s exponentially increased the numbers of subscribers that could be served by each operator, raising the number from fifty to one hundred, depending on the type of switchboard used, to as many as ten thousand. Consequently, it became impossible for operators to learn the names of each subscriber served, and this instance of "distancing" was "one of the earliest retreats from personalized service" on the path to fully automated dialing.[41] Another result of the upgrade to multiple switchboards was that operators began connecting calls by number, rather than by name. The telephoning public initially resisted telephone numbers, nearly as much as dialing years later. By the time the dial was introduced, however, callers had already accepted the responsibility of remembering or retrieving each other's telephone numbers. The introduction of telephone numbers was a pivotal step on the way toward Bell convincing its customers to begin dialing. Once callers accepted the use of numbers to identify each other's telephones, it became much easier to convince them to begin dialing those numbers instead of saying them to an operator.

The shift from names to numbers as the means of telephonic identification began in the late 1880s, yet after the turn of the century customers were still complaining. A 1902 editorial in the *Boston Post*, for instance, criticized the New England Telephone and Telegraph Company for its rule of connecting all calls by number only, arguing instead that local operators should continue to connect their more common requests by name. For customers that operators were still personally familiar with, the paper insisted, it would be easier for "an operator to make the connection than to waste time arguing with the caller" about which mode of identification to use.

Resting on an argument about the efficiency of common sense, the *Boston Post*'s editorial predicted that "the rule will break down of its own weight." Other instances of aversion to telephone numbers, meanwhile, were triggered by class-based resentments. As late as 1911, for example, a lawyer in Greenville, South Carolina, named D. Lewis Dorroh sued the Southern Bell Telephone and Telegraph Company for $500, citing the company's "refusal to fulfill its obligation to connect him without any work on his part." The work in question for Dorroh was not dialing, still more than a decade away for Greenville residents, but remembering or retrieving the telephone number of the party he wanted to reach (which, according to the facts of the case, was his country club).[42]

Alexander Graham Bell was awarded the first U.S. patent for telephone technology in 1876, and that year the company named for him began to sell subscriptions to their new service. Bell himself may have telephonically beckoned to Watson directly, without a human mediator between them ("Mr. Watson, come here. I want you."), but by the time his invention was being marketed to the public, speaking to someone on a Bell telephone meant speaking first to an operator. The presentation of operators as pliant, attentive servants was a substantial feature of Bell's marketing in the telephone's formative era, meant to distinguish Bell subscribers from the rest. By contrast telephone numbers had a sort of leveling effect among telephones and, by extension, their users. Now the Greenville, South Carolina, country club of D. Lewis Dorroh, Esq. was subject to the same identifier as any shop with a phone. Likewise, all residences were reduced to numbers regardless of the inhabitants' wealth, prestige or power. Identifying telephones by number instead of by name made it more difficult to recognize the social status of the person using a particular phone.

No less than dialing, telephone numbers were initially resisted and eventually accepted. Over time many people embraced their telephone numbers, and some even grew attached to them. Since their introduction over one hundred years ago, most systematic changes to telephone numbers have met with protest. The easy addition of area codes was an exception to the rule. When area codes were introduced, personalized service was no longer an expected feature of telephone service; but, ironically, area codes reintroduced a sense of status and social identity to telephone numbers. This accrual of status happened gradually over the second half of the twentieth century, as geographical places became indexed to their area codes. (Rappers, for example, use area codes as shorthand for identifying roots and home turf.) Meanwhile, as the twentieth century waned, almost fifty years after the introduction of area codes and more than one hundred years after the introduction of telephone numbers, telecommunications companies in the U.S. began to claim that they were running out of room. To accommodate the demand for cellular phones, pagers, faxes, modems and dial-up connections in silicon alleys and valleys up and down both coasts, all of which required a telephone number, many telecom companies implemented a series of local "splits" and "overlays," whereby major metropolitan areas were divided into two or three area codes, or new area codes were woven into existing ones, making available twice as many possible telephone numbers. For example, in Chicago telecom companies "split" the city and assigned the North Side a new 773 area code, while the downtown "Loop" and the South Side remained 312. In New York City, on the other hand, a 646 area code was "overlayed" onto Manhattan's traditional 212. The extent of personal attachments to area codes would be revealed when the splits and overlays threatened some callers with the prospect of losing theirs.

Local newspapers editorialized against the inconvenience and impersonality of local calls that require eleven digits (1+area code) instead of a cozier seven. The *San Francisco Examiner* criticized the system of proposed Bay Area overlays as "maddening," and implored is readers to act: "we must get rid of [them]." The *St. Petersburg* (FL) *Times* warned readers that the extra digits were "only the beginning," and quoted Ronald Connors, head of the North American Numbering Plan Administration, who warned that by 2025 all possible area codes could be in use, and the result may be adding another digit to everyone's telephone number. The frustration in both of these examples is overtly focused on the added time and memory the overlays require for local calls, although such reactions were also, arguably, animated by the loss of distinction between calls next door and across the country. *Seinfeld*, the popular '90s sitcom, suggested that the status-savvy urbanites had identified another problem with the splits and overlays. During a 1998 episode, Elaine's new 646 area code for her home phone brought enough "social ostracization" that she tried to persuade a fatally ill neighbor to hand over her 212 number. "Moving" to a new telephonic outpost made people like Elaine anxious because it meant losing the status that accrues to a Manhattan residence, identifiable through a 212 area code, as opposed to one of the outer boroughs which share 718, a less-impressive lot of telephone real estate. The subsequent proliferation of overlays for cellular service may have "divorced [area codes] from their areas," and relaxed the residential stratification reflected in telephone numbers, but in its coverage of these cellular overlays, the *New York Times* insisted that residential area codes remained "an aspect of one's identity, even a status symbol, such as signifying a long-time resident" of a reputable or exclusive part of town.[43]

While newspapers opposed the new areas codes and sitcom characters fretted and fought over them, a proposed West Side overlay in Los Angeles met with organized resistance. A telephone-owning plastic surgeon named Steven Teitelbaum responded to what he felt was a "condescending" notice from the phone company, alerting him to the new overlay and subsequent dialing burden, by writing "bull" across the top and faxing it to his friend Robert Scheer, a nationally syndicated newspaper columnist. Scheer also happened to be a former engineering student who knew enough about telecom infrastructure (and business practices) to smell a rat. Scheer began conducting research and interviews, and soon he concluded that the "only shortage we ever had was one of truth," not available telephone numbers. Only 33 million of the available 180 million telephone numbers in California were being used, and telephone companies were "hoarding" unused numbers and pushing for the new area codes, rather than fixing an inefficient system for assigning unused numbers and reassigning abandoned ones, or overlaying technologically specific area codes for faxes, modems and ATMs.[44]

The "Stop the Overlay" campaign found a publicity machine in Scheer, who wrote regularly about the campaign in *Our Times*, the West Side weekly version of *The Los Angeles Times*, for which Scheer also wrote, but whose editors refused to cover the overlays in their daily edition. Organizers also launched a website, www.stopoverlay.com (now defunct), which attracted over 9000 hits in May, 1999, due in no small part to Scheer's promotion of it. Scheer's muckraking and a subsequent letter, fax and email campaign to the F.C.C. (who oversaw CPUC, the California Public Utility Commission, which managed telephone numbers in the state) inspired some local hearings on the proposed overlay. In turn, California State Assemblyman Wally Knox, with support from fellow West Side Democrat, U.S. Representative Harry Waxman, introduced Assembly Bill 818. The bill soon passed and became known as the 1999 Consumer Area Code Relief Act, which required telecom companies to dialogue with the public before implementing changes to area code assignments. Not surprisingly, however, the telecom industry's lobbying was more sustained and ultimately more successful that an ad hoc movement to prevent one particular overlay. In response to pressure at the federal level, CPUC proposed to split the West Side into two area codes, rather than the overlay originally proposed. One result of the new proposal was to literally divide the opposition, specifically to dampen resistance among those West Siders who would get to keep their 310 area code rather than being reassigned to 424. The shift from an overlay to a split "did not silence opposition to the new area code change, but it was not met with action comparable to the 'Stop the Overlay' campaign. The telephone companies were then able to achieve their objectives over a more limited and less organized resistance." Finally, in August 2005, the CPUC "unanimously approved a plan for an all-services area code overlay to the existing 310 area code in Southern California," with little public opposition (and no coverage by Scheer).[45]

The "Stop the Overlay" campaign was reminiscent of reactions, nearly forty years earlier, to the "last radical change in telephone numbers," the switch from alpha-numeric exchange names to all-digit telephone numbers. The shift to all-digit dialing meant that the telephone exchange for Manhattan's Upper East Side, for example, morphed from BUtterfield-8, made famous by the John O'Hara novel and the film starring Elizabeth Taylor, into the less-glamorous 288 (ironically, also the code for AT&T). Even in neighborhoods less exclusive than BUtterfield-8, exchange names became "a source of pride (and snobbery) among the city's residents" as well as a source of pride (if not snobbery) for residential telephone subscribers across the country.[46] Four decades later, the class mappings made possible through area codes were explicitly a source of humor in the *Seinfeld* episode spoofing the Manhattan overlay, and they also animated resistance during the "Stop the Overlay" campaign. During the 1960s, when the problem was disappearing letters rather than proliferating area codes, resistance to all-digit dialing was mobilized against the unilateral arrogance of the telecom industry and in defense of individual rights and consumers' quality of life. Like the Stop the Overlay campaign, opposition to all-digit dialing sprang up in California, where the Anti-Digit Dialing League was founded in San Francisco, during June, 1962. The ADDL's founder was longtime Bay Area activist Carl V. May, but their public leader was S. I. Hayakawa, a prominent linguist at San Francisco State University who did most of the public speaking for the group and coined the phrase "creeping numeralism" to criticize the retirement of exchange names and all it presaged. (Hayakawa would achieve greater notoriety only a few years later, as an SFSU administrator publicly and violently opposed to student protests on campus.) The first national notice of the ADDL appeared in the *New York Times*, less than a month after the league's founding. It is worth noting that the first mention of the anti-digit dialers in the *Times* was also the paper of record's first mention of all-digit dialing itself. As with Scheer and the STO campaign, a basic "victory" for the ADDL was making telephone companies account for a service change before just going ahead and making it.[47]

AT&T's attempt to impose all-digit dialing without soliciting acceptance from its customers was not simply the result of its established monopoly status. While the company did send out "Dear Subscriber" notices, I have not found any promotional (or instructional) films produced by Bell for all-digit dialing, unlike earlier changes to telephone use, such as dialing for local and then long-distance calls, or even for intermediate stages of automation such as "operator toll dialing." By the time exchange names were retired in favor of all-digit dialing, the protocols of telephone use in the U.S. had absorbed more than just remembering and dialing telephone numbers. On top of these experiential naturalizations, Bell had successfully groomed the dialing public to expect that changes in telephone service were technological solutions to problems caused by continually growing demand. During the late 1990s, newspapers (and sitcoms) noticed and were wary of the area code splits and overlays, but it wasn't until Scheer's investigative journalism and the STO campaign that the overlays were characterized as anything but inevitable. Likewise, when "288" (AT&T) announced that it was dropping exchange names in favor of digits, almost forty years earlier, newspapers lamented this change while accepting it as inevitable. For instance, a December, 1962 editorial in the *Christian Science Monitor* accepted the shift as unavoidable. "Telephone company officials say they would like to keep exchange names but that growth of telephone service simply requires more numbers than are possible with letter combinations." All-digit dialing would increase possible telephone numbers by about half the present amount, the editorial reports, making it "hard to argue" with the change. Similarly, three days after "breaking" all-digit dialing (and the ADDL) as a news story, the *New York Times* ran an editorial acknowledging that "Butterfield, Magnolia, Sweetbriar, Cantaloupe and all of the beautiful exchanges of yesterday [are g]one or going perhaps with the buffalo and the steam locomotive. We wish they wouldn't." The *Times,* like the *Christian Science Monitor*, found the change inevitable even while announcing the formation of the ADDL and wishing "more power to them."[48]

The *Times* ran two separate editorials about all digit-dialing; a second one focused on the ADDL, after one had already bade "farewell [to] Magnolia 4" (its title) and the other "beautiful exchanges of yesterday." The first editorial asks, "Is this the result of automation: more work for the consumers, less for the company?" For Robert Scheer and STO, the latter complaint was their more pressing concern; before the telephone company imposed new numbers and systems for distributing them, Scheer and company insisted that the companies must first prove any proposed changes necessary and then cooperate with the public while implementing them. The splits and overlays were less the stakes of their fight, ultimately, than the power of telephone companies. For the ADDL, more numbers were precisely the problem; indeed, "creeping numeralism" became their sound byte battle cry. "Strain on the brain" was another ADDL slogan. (Shudder to think how they would have reacted to local calls requiring area codes!) Like the STO campaign (and editorials against the splits and overlays), the Anti-Digit Dialers were worried about the added memory necessary to remember numbers instead of letters, and the enemies in both instances were the same, phone companies and regulatory agencies. The ADDL was more focused on the experience of telephony itself, specifically the relative inhumanity of identifying ourselves telephonically with numbers instead of names.

The *Times* editorial about all-digit dialing did not answer its own question about automation and added work for consumers. Instead, another question is immediately raised, one perhaps even closer to the heart of the matter: "why can't we have romance in the telephone business they way we used to?" Here we may glimpse a shadow of "the voice with a smile" as the true prelapsarian paradise animating nostalgia for exchange names. News briefs about all-digit dialing and editorials lamenting it were fairly common during late-1962 and 1963, and occasionally feature articles about the phenomenon and resistance to it were written as well. One such article, published in the *Los Angeles Times* in March 1964, begins by linking all-digit dialing to "social security numbers, license plate numbers, zip-code numbers and bank-account numbers," and asks "does it sometimes seem to you that you're losing your identity as a name and drowning in a sea of numbers?" The ADDL are then described as "a ray of hope" from upstate in San Francisco and a "noble movement" resistance the U.S.'s devolution into "a numerical society." (The previous chapter cited Daniel Boorstin, who claimed the cash register helped complete this transformation during the late-nineteenth century.) Operator service is never mentioned in the article, its disappearance having been accepted prior to the retirement of exchange names. However, an accompanying cartoon hints at a lingering connection between exchange names and operators. As a group of exasperated callers

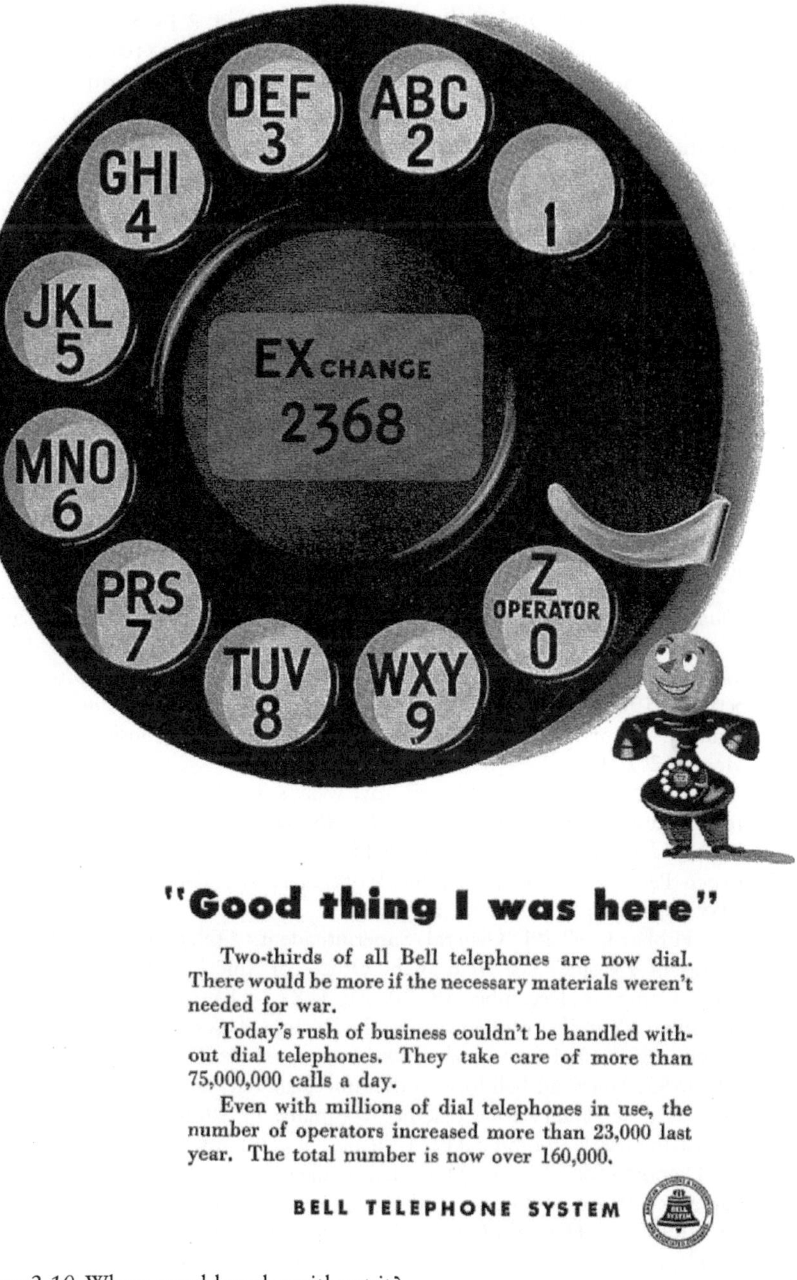
DEF 3
ABC 2
GHI 4
1
JKL 5
EXCHANGE
2368
MNO 6
PRS 7
Z OPERATOR 0
TUV 8
WXY 9
"Good thing I was here"
Two-thirds of all Bell telephones are now dial. There would be more if the necessary materials weren't needed for war.
Today's rush of business couldn't be handled without dial telephones. They take care of more than 75,000,000 calls a day.
Even with millions of dial telephones in use, the number of operators increased more than 23,000 last year. The total number is now over 160,000.
BELL TELEPHONE SYSTEM

*Figure 2.10* Where would we be without it?

climb over a giant phone, drowning in an avalanche of numbers, the caption asks: "when you're dialing 1-212-797-5500 [as opposed to SWeetbriar-7], don't you long for that helpful voice?" The caption suggests that the addition of extra numbers to store and remember made some callers long for the days of operators on call and ready to serve.[49]

The ADDL understood dialing a telephone to be a personal, social and emotional experience, above and beyond its more familiar role as a medium for such experiences. Added memory and "strain on the brain" were concerns of the ADDL, but added work for telephone callers faded as a concern, less significant than "creeping numeralism." The removal of exchange names was justified to a skeptical public as necessary to accommodate increasing demand, as per usual; however, all-digit dialing was technologically necessary to pave the way for digital telecommunications infrastructure. When telephony underwent digitization, dialing was already established as an everyday, if overlooked, technological assemblage. All-digit dialing not only removed letters from telephone numbers, it also marked a turning point in the history of telephony: automation was complete and digitization began. Once callers began "dialing" on touch-tone keypads, as the next chapter describes, the "creeping numeralism" of all-digit dialing would begin to mutate and metastasize into digital interface.

## Notes

1. *Phantom of the Operator* is an "ephemeral film," in which director Caroline Martel "takes overlooked artefacts of cinema history – one hundred industrial, advertising and scientific management films produced in North America between 1903 and 1989 – and turns them into a dreamlike montage documentary." http://artifactproductions.ca/fantome/en/film/synopsis.htm. Accessed March 12, 2015.
2. Sidney H. Aronson, "Bell's Electric Toy: What's the Use? The Sociology of Early Telephone Usage," *The Social Impact of the Telephone*, Ithiel de Sola Pool, ed. (Cambridge, MA: MIT Press, 1977): 15–39.
3. Steven Peter Vallas, *Power in the Workplace: The Politics of Production at AT&T* (Albany: SUNY Press, 1993): 37.
4. A.S. Hibbard, AT&T General Superintendent, "Telephone Switch Boards: Report of a Conference Held at the Offices of American Telephone and Telegraph Company, December 19, 20, 21, 1887," 114, quoted in Venus Green, "Goodbye Central: Automation and the Decline of 'Personal Service' in the Bell System, 1878–1921," *Technology and Culture* 36 (1995): 921.

    In 1885, American Bell formed the American Telephone and Telegraph Company as a subsidiary long distance company. In 1900, the directors of Bell American made a booming AT&T the parent company. In this chapter I refer to "Bell" and "AT&T" more or less interchangeably.
5. For accounts of Strowger as the origin of automation, see, for example: J.E. Kingsbury, *The Telephone and Telephone Exchanges: Their Invention and Development* (New York: Longsman, Green and Co., 1915); Harry B. MacMeal,

*The Story of Independent Telephony* (Chicago: Independent Pioneer Telephone Association, 1934); Marion May Dilts, *The Telephone in a Changing World* (New York: Longsman, Green and Co., 1941); M.D. Fagen, ed., *A History of Engineering and Science in the Bell System: The Early Years, 1875–1925* (Bell Telephone Laboratories, 1975).

6. American Bell Telephone Company patent attorney Thomas D. Lockwood to Bell President John E. Hudson, 11 April 1891, AT&T Archives Box 1286, quoted in Venus Green, *Race on the Line: Gender, Labor, and Technology in the Bell System, 1880–1980* (Durham, NC: Duke University Press, 2001): 118.

7. MacMeal, *The Story of Independent Telephony*, 6.

8. Amy Sue Bix, *Inventing Ourselves Out of Jobs? America's Debate over Technological Unemployment, 1929–1981* (Johns Hopkins UP, 2000), 3.

9. Bix, *Inventing Ourselves out of Jobs?*, 1.

10. Ibid., 5–6.

11. Venus Green, *Race on the Line*; Steven H. Norwood, *Labor's Flaming Youth: Telephone Operators and Worker Militancy, 1878–1923* (Urbana, IL: University of Illinois Press, 1990).

12. Susan Porter Benson, *Counter Cultures: Saleswomen, Managers and Customers in American Department Stores, 1890–1940* (Urbana, IL: University of Illinois Press, 1986).

13. "Dial Phones, Banned in Senate, Stir House," *Washington Post*, May 23, 1930, 1; "Roll Your Own Phones Keep Users Arguing," *Washington Post*, May 24, 1930, 1; quoted in Green, *Race on the Line*, 308n.2.

14. Unlike cars, electricity, and household appliances, telephone saturation dipped during the Great Depression, and it would not be until after World War Two that phones would again become as common as cars (and they did not catch the radio until the introduction of cell phones). Claude Fischer, "Figure 1: U.S. Households with Selected Consumer Goods," *America Calling: A Social History of the Telephone* (U California P, 1992), p. 22.

15. Green, "Goodbye Central," 917, 925.

16. Green, *Race on the Line*, 119–20.

17. Milton Mueller, "The Switchboard Problem: Scale, Signaling and Organization in Manual Telephone Switching, 1877–1897," *Technology and Culture* 30 (1989): 534–60.

18. Quoted in Green, *Race on the Line*, 120.

19. Southern New England Telephone Company, "Brief: The Automatic System in Connecticut," 1904, p. 1; quoted in Green, *Race on the Line*, pp. 120–1, 299n19.

20. Print ad in possession of author.

21. Green, *Race on the Line*, 122.

22. Green, "Goodbye Central," 918; Katherine Schmitt, "I Was Your Old Hello Girl," *Saturday Evening Post*, July 12, 1930, 20.

23. Norwood, *Labor's Flaming Youth*, 33.

24. Julia O'Connor, "History of the Organized Telephone Operators' Movement," Part 4, *Union Telephone Operator*, May, 1921, 15, quoted in Norwood, 158.

25. Norwood, 169.

26. Print ad in possession of author.

27. Green, "Goodbye Central," 915–16.

28. Bix, 159.

29. Green, *Race on the Line,* 127.
30. Ibid., 160.
31. Bix, 91.
32. Ethel Best, "The Change from Manual to Dial Operation in the Telephone Industry," U.S. Department of Labor, Bulletin of the Women's Bureau, 110 (1933): 2, 9.
33. Green, *Race on the Line,* 162.
34. *How to Use a Dial Phone,* http://www.archive.org/details/HowtoUse1927. Accessed February 29, 2016. Screenshots by author.
35. Print ad in possession of author.
36. On telephone use in rural communities, see Lana F. Rakow, *Gender on the Line: Women, the Telephone and Community Life* (Urbana, IL: University of Illinois Press, 1992).
37. *Dial Comes to Town.* http://www.archive.org/details/DialComesToT. Accessed February 29, 2016. Screenshot by author.
38. Green, *Race on the Line,* 163.
39. Bix, 224.
40. *The Nation at your Fingertips,* http://www.archive.org/details/Nationat1951. Accessed February 29, 2016.
41. Green, "Goodbye Central," p. 923.
42. General Counsel H. E. W. Palmer, Southern Bell Telephone and Telegraph Company to John H. Peck, Esq., Legal Department, AT&T, April 11, 1911, quoted in Green, "Goodbye Central," 924–5. Dorroh lost his case.
43. "A New Plan of 'Overlays' for Bay Area Phone Numbers is Cumbersome, Annoying, Confusing and Totally Unnecessary," *San Francisco Examiner,* May 3, 1999; Howard Troxler, "Don't Get Hung Up on Extra Phone Digits; It's Only the Beginning," *St. Petersburg Times,* December 12, 1997; Simon Romero, "Now You Need an Area Code Just to Call Your Neighbors," *New York Times,* May 7, 2001; Ian Urbina, "Area Codes, Now Divorced From Their Areas," *New York Times,* October 1, 2004.
    The New York State Public Service Commission petitioned the Federal Communications Commission in an attempt to stop a New York City overlay; in its denial, the U.S. 2nd Circuit Court acknowledged that "the apprehension about the addition of new area codes (and the concomitant addition of numbers required to dial a local call) has reached levels that the controversy has found its way into popular culture." This *Seinfeld* episode is cited as evidence, along with a 2000 *Simpsons* episode, during which Homer runs for mayor of "New Springfield," after an area code split divides the community (People of the State v F.C.C., docket number 99-4205, n1).
44. Wan-Ying Lin and William H. Dutton, "The 'Net' Effect in Politics: The 'Stop the Overlay' Campaign in Los Angeles," *Party Politics* 9 (2003): 124–136; Robert Scheer, "Only shortage we ever had was one of truth," *Los Angeles Times,* March 19, 2000.
    Lin and Dutton's account of the "Stop the Overlay" campaign adopts it as a case study in order to analyze the efficacy of web-based consumer and community activism. It is worth noting that the authors remain silent on the social, residential and professional, rather than technological, "networks" activated among Angelinos like Scheer and Teitelbaum in order to mobilize resistance to the overlay.
45. Lin and Dutton, p. 130; CPUC Press Release, "PUC Approves All-Services Overlay for 310 Area Code," August 25, 2005.

46. Simon Romero, "Now You Need an Area Code Just to Call Your Neighbors," *New York Times*, May 7, 2001.
47. Lawrence Davis, "All-Digit Dialing Fought on Coast," *New York Times*, June 24, 1962.
48. *Christian Science Monitor*, "Phone Calls – 'By the Numbers,'", December 27, 1962; "Magnolia 4, Farewell," *New York Times*, June 27, 1962; "The Anti-Digit Dialers," *New York Times*, October 2, 1962.
49. Richard Gehman, "Number, Please!," *Los Angeles Times*, March 29, 1964.

## Bibliography

"The Anti-Digit Dialers," *New York Times*, editorial, October 2, 1962.

Aronson, Sidney H. "Bell's Electric Toy: What's the Use? The Sociology of Early Telephone Usage." *The Social Impact of the Telephone*, Ithiel de Sola Pool, ed. Cambridge, MA: MIT Press, 1977: 15–39.

AT&T, "How to Use a Dial Phone," 1927 http://www.archive.org/details/Howto Use1927. Accessed April 6, 2016.

AT&T, "Dial Comes to Town," circa 1950 http://www.archive.org/details/Dial ComesToT. Accessed April 6, 2016.

AT&T, "Speeding Speech," (Sound Productions, circa 1950s) https://archive.org/ details/Speeding1950. Accessed April 6, 2016.

AT&T, "The Nation at Your Fingertips," (Audio Productions, 1951) http://www. archive.org/details/Nationat1951. Accessed April 6, 2016.

Benson, Susan Porter. *Counter Cultures: Saleswomen, Managers and Customers in American Department Stores, 1890–1940*. Urbana, IL: University of Illinois Press, 1986.

Best, Ethel. "The Change from Manual to Dial Operation in the Telephone Industry." U.S. Department of Labor, Bulletin of the Women's Bureau 110 (1933).

Bix, Amy Sue. *Inventing Ourselves Out of Jobs? America's Debate over Technological Unemployment, 1929–1981*. Baltimore: Johns Hopkins, 2000.

Davis, Lawrence. "All-Digit Dialing Fought on Coast." *New York Times*, June 24, 1962.

Dilts, Marion May. *The Telephone in a Changing World*. New York: Longsman, Green and Co., 1941.

Fischer, Claude. *America Calling: A Social History of the Telephone*. Berkeley: University of California Press, 1992.

Gehman, Richard. "Number, Please!" *Los Angeles Times*, March 29, 1964.

Green, Venus. "Goodbye, Central: Automation and the Decline of 'Personal Service' in the Bell System, 1878–1921." *Technology and Culture* 36 (1995): 912–49.

Green, Venus. *Race on the Line: Gender, Labor and Technology in the Bell System, 1880–1980*. Durham, NC: Duke UP, 2004.

John, Richard. *Network Nation: Inventing American Telecommunication*. Cambridge, MA: Belknap Press, 2010.

Lin, Wan-Ying and William H. Dutton. "The 'Net' Effect in Politics: The 'Stop the Overlay' Campaign in Los Angeles." *Party Politics* 9 (2003): 124–36.

MacMeal, Harry B. *The Story of Independent Telephony*. Chicago: Independent Pioneer Telephone Association, 1934.

"Magnolia 4, Farewell." *New York Times*, June 27, 1962.

Mueller, Milton. "The Switchboard Problem: Scale, Signaling and Organization in Manual Telephone Switching, 1877–1897." *Technology and Culture* 30 (1989): 534–60.

"A New Plan of 'Overlays' for Bay Area Phone Numbers is Cumbersome, Annoying, Confusing and Totally Unnecessary." *San Francisco Examiner*, May 3, 1999.

Norwood, Steven H. *Labor's Flaming Youth: Telephone Operators and Worker Militancy, 1878–1923*. Urbana, IL: University of Illinois Press, 1990.

*Phantom of the Operator*. Directed by Caroline Martel. Montreal: Artifact Productions, 2004.

"Phone Calls – 'By the Numbers,'" *Christian Science Monitor*, December 27, 1962.

Rakow, Lana F. *Gender on the Line: Women, the Telephone and Community Life*. Urbana, IL: University of Illinois Press, 1992.

Romero, Simon. "Now You Need an Area Code Just to Call Your Neighbors." *New York Times*, May 7, 2001.

Scheer, Robert. "Only shortage we ever had was one of truth." *Los Angeles Times*, March 19, 2000.

Schmitt, Katherine. "I Was Your Old Hello Girl." *Saturday Evening Post*, July 12, 1930.

Troxler, Howard. "Don't Get Hung Up on Extra Phone Digits; It's Only the Beginning." *St. Petersburg Times*, December 12, 1997.

Urbina, Ian. "Area Codes, Now Divorced From Their Areas." *New York Times*, October 1, 2004.

Vallas, Steven Peter. *Power in the Workplace: The Politics of Production at AT&T*. Albany: SUNY Press, 1993.

# 3    Then Press #
## Touch-Tone Phones and
## Digital Interface

Fade in. Dateline, 1971. A rotary dial and a touch-tone keypad, unattached to telephones, mounted like modern art in the white cube. A dial tone breaks the silence, conjuring two hands. In under five seconds a manicured white male hand presses seven buttons on the keypad, then eight more seconds pass while an identical hand continues to turn and release the dial, waiting each time for it to return to its resting position before turning it again. Once the dialer finishes, a narrator announces in mock-solemn baritone: "It's the greatest thing since the wheel: touch-tone calling." To promote rotary dialing AT&T had produced instructional films lasting twenty minutes; this commercial justified the upgrade to touch-tone calling with a quick demo and sarcastic tagline. The visual presentation is straightforward and to the point: pushing buttons on a touch-tone keypad is faster than rotary dialing. The ad's tenor, meanwhile, demonstrates how AT&T wanted its customers to approach the new interface. Describing touch-tone calling as "the greatest thing since the wheel" is a play on the dial, and touch-tone was the telephone's first makeover since callers began dialing the rotary "wheel." But the narrator's tone is dismissive rather than celebratory, signaling to viewers that, actually, touch-tone is no big deal.

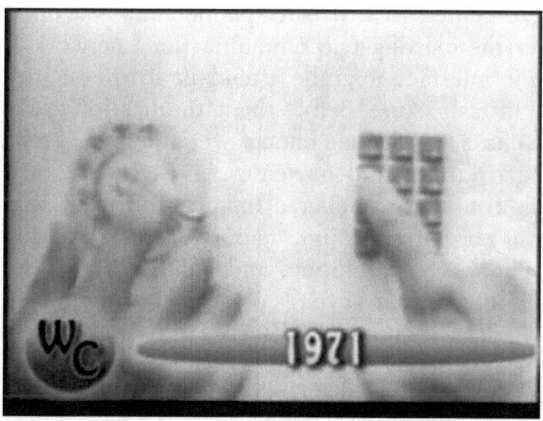

*Figure 3.1* "The greatest thing since the wheel: touch-tone calling."[1]

Cut to a low stage in front of a brick wall. Dateline, about forty years later. A balding man with a red goatee holds a microphone, into which he is complaining about dialing a telephone. His audience chortles at the reminder of the time and effort dialing required, before the once-routine procedure was rendered absurd to contemporary sensibilities. "We used to hate people with zeros in their numbers. 'A nine *and* a zero? How bad do I want to talk to this asshole?'" The joke comes from a bit by Louie CK about people taking cell phones for granted, and it echoes AT&T's touch-tone commercial: both proffer admiration for state-of-the art telephone technology, however ironically, by mocking the slowness and relative burden of rotary dialing.[2] Dialing introduced a drastic change into the basic experience of telephony, and AT&T's marketers and managers knew their subscribers would recognize it as an imposition. The previous chapter detailed how, for decades, AT&T promoted the dial to subscribers as an improvement rather than a reduction in telephone service. Pioneering self-service grocers undertook similar campaigns for their stores promising a better shopping experience as well as lower prices. Touch-tone calling, conversely, required no marketing campaign to win over the dialing public. Trading in a rotary phone for a new touch-tone model did not eliminate the presence in one's life of an iconic, everyday service employee. Furthermore, customers were not forced to upgrade when the telephone company said so, as they had been with the dial. Touch-tone was not foisted on individual telephone customers as directly or immediately as the dial had been; there was no "cutover" to touch-tone. One pithy commercial, released almost a decade after touch-tone phones hit the market, marked the apogee of AT&T's national marketing for the telephone keypad.

Alongside clocks and calculators, telephones were among the first everyday consumer technologies to undergo digitization, and "touch-tone calling" provides an origin story for the history of digital interface. AT&T spent the sixties and seventies adapting its national networks to accommodate traffic between computers as well as telephones, and "the internet itself was largely laid over the existing telecommunications network."[3] Touch-tone keypads were the interface upgrade attendant to the digitization of telecommunication infrastructure. Before the naturalization of keyboards, cursors and mouse clicks, touch-tone phones were the first everyday consumer technology used to access and navigate digitized networks. Many tasks and transactions routinely conducted online, from checking the weather to managing your checking account, migrated to personal computers from touch-tone telephones. Smart phones and hand-held gadgets now feature interactive touch screens, but keypads continue to linger within a range of digital assemblages, the first of which were ATMs, subject of the next and final chapter. Meanwhile, whenever the time comes to enter numerical data into a smart phone, a pixelated keypad appears on the screen.

For American consumers, the first outright purchase of telecom technology *en masse* was not the cell phone or personal computer, but the touch-tone telephone. AT&T rolled out touch-tone in 1963, yet rotary models remained a statistical majority in the U.S. into the 1980s. The break up of AT&T's monopoly ushered in the keypad as everyday interface in the U.S. AT&T had always leased telephones to its subscribers (through Western Electric and other manufacturing subsidiaries), so after divestiture consumers encountered a new market of residential telephones for sale. For two decades AT&T's subscribers had been free to choose when (or whether) to upgrade to a touch-tone model, but the new market in home telephones pushed the dial toward extinction. Many manufacturers and retail outlets offered rotary phones, but virtually everyone purchased touch-tone models. Beyond faster dialing, two other new markets contributed to touch-tone's takeover. First and foremost, the break-up of AT&T's monopoly opened long-distance telephone service to competition. The digital networks of long-distance upstarts, such as MCI and Sprint, could not be accessed via the electronic pulses generated by turning a rotary dial; only by using a touch-tone phone could consumers capitalize on the savings offered by AT&T's new long-distance competitors. The functional retirement of rotary dialing also spurred a new industry in "audiotext" information services, such as weather reports, stock tickers, and sports scores, which callers could only navigate via touch-tone phones.

The "dial era" lasted roughly forty years, from 1920 to 1960, while two landmarks of federal policy bookend the touch-tone era in the 1980s and 1990s.[4] The divestiture of AT&T's monopoly was "the most spectacular by-product of [the] transition to a neoliberal development policy" for tele-communications in the U.S., and there are several scholarly analyses of it.[5] Focused on the break-up's commercial, corporate and financial implications, these studies overlook the central role that touch-tone played for consumers during the aftermath.[6] If AT&T's divestiture ushered in the touch-tone era, then the Telecommunications Act of 1996 brought the curtain down. Scarcely a decade after divestiture, the Act was a further milestone of deregulation. Touch-tone demanded its fair share of attention, although the legislation helps mark the end of the touch-tone era because keypads (and telephones) were overshadowed by unprecedented deregulation of telecommunication infrastructure, especially those networks beginning to comprise the Internet.[7] The touch-tone era can also be said to have ended during the late 1990s, in another sense, when cell phones assumed center stage within telecommunications policy and consumer technology markets.

The previous chapter presented a (consumer) labor history of rotary dialing, focused on callers instead of operators; this chapter adopts a cultural (policy) studies approach to present the first critical history of telephone keypads. After telling an origin story for the keypad in the first section, I devote one section apiece to the three new markets that were opened in the wake of AT&T's divestiture: home telephones, long-distance calling, and "audiotext" information services. In the absence of any critical scholarship devoted to the telephone keypad, my archive for this chapter consists of the policy documents, legislation, lawsuits, and Supreme Court rulings that governed the emergence of touch-tone phones and digital dialing. At the state and especially federal level, these laws, policies and rulings helped set the experiential as well as infrastructural stage for the internet. "Touch-tone calling" may already sound as quaint as dialing, so in order to capture how the dialing public responded to touch-tone phones and new forms of digital interface, I also analyze newspaper advice columns in the wake of AT&T's break-up, as well as op-eds and letters to the editor.

Personal computers and smart phones have absorbed much of touch-tone's functionality, thanks to keyboards, mice, and more recently touchscreens that exceed the keypad's versatility. In a recent article elaborating "a labour theory of the iPhone era," Jack Linchuan Qiu, Melissa Gregg and Kate Crawford forge a welcome critique of digital labor (one of the "new labors" critiqued in the introduction) as a category "whose delineation is above all technological." The history of dialing, and especially its afterlife as keypad data entry, emphasizes how much digital labor studies stands to gain from emphasizing human digits. Laboring digits, namely index fingers, began turning rotary dials decades before telecommunications went digital in the more familiar sense. In a similar vein, Lisa Nakamura has described using an iPhone as "digital manual labor."[8] Critics since Marx have railed against the deracination of labor into abstract units such as factory- and farmhands, while sayings like "all hands on deck" and "punch the clock" connote the blood, sweat, and tears of physical toil. Digital labor must at some level entail bodies as well as machines, but the potential of digital labor as a critical category will be compromised if an orientation toward human digits reduces the term into another synecdoche for the drudgery of work – from hands to fingers – updated now for life after industrialism. To avoid repeating the abstraction of working bodies into laboring body parts (first as tragic hands, now as farcical digits), in this chapter I follow Nakamura's corporeal turn for digital labor while also opening the purview for digital labor historically.

Qiu, Gregg, and Crawford's foregrounding of FoxConn (as opposed to Facebook) within the iPhone's "circuits of labor" is a welcome corrective highlighting the "material production of immaterial culture." I hope this chapter might similarly help digital labor studies begin to focus more on the touchy embodiments of technological interface. Within Qui, Gregg, and Crawford's cultural studies approach to digital labor, what we might call the manual digital labor at FoxConn is given equal billing with Nakamura's "digital manual labor" of iPhone users. iPhones feature touch screens and are categorized as "smart;" however, zeros and ones remain less productive digits historically than fingers and thumbs. The laboring digits at FoxConn belong to employees, while the digital productivity of iPhone labor extends to consumers. Tiziana Terranova appropriated the term "free labor" to describe unpaid contributions of Internet users to online content and software. The burgeoning digital labor archive already swells with analyses of user-generated content online, and the smart phone has become an assemblage for digital leisure as well as labor. Yet so much unpaid digital labor, what Terranova called free labor, is boring or even tedious, and certainly not all of it is fun. Profits and productivity are wrung from ATM withdrawals as well as from tweets, while Twitter can be described as an example of creative free labor being digitized into a more productive routine.

In describing the use of an iPhone as digital manual labor, Nakamura sought to capture how quickly what she calls "the boom! moment" of new consumer technology wears off. Even on an iPhone, "the manual labor of interface manipulation becomes laborious soon enough, just like all the other interface required of us for work and entertainment."[9] The novelty of consumer technologies like the iPhone gives way to banality when their "booms" are forgotten. Most transformative new media change the world more as they age, not less, and it is "when technologies such as the telephone and the computer cease to be sublime icons of mythology and enter the prosaic world of banality that they become important forces of social change."[10] The history of digital interface demonstrates that this is the case not only for the telephone, but also for the telephone dial. The previous chapter described how the rotary telephone dial's "boom! moment" lasted decades. The telephone keypad never really experienced a "boom! moment" in the first place, then became an assemblage far more powerful than the dial ever was.

## Pushing Buttons

At *SxSW Interactive* in 2010, Bill DeRouchey presented a "history of the button." The first push buttons were found on flashlights and cameras, beginning in the 1890s. Next came doorbells and light switches, before radio preset buttons initiated the now-familiar practice of saving favorite settings. DeRouchey argues that the lever was the signature technology of the mechanical age, while the button became the iconic interface of the electronic age. The act of pushing a button was also an unprecedented form of "abstracted motion." Unlike pulling a lever, a pressed button bore no mechanical relationship to the reaction it generated, whether the ringing of a bell, the snap of a shutter or the illumination of a room. During the 1950s and 1960s, buttons connoted, among other things, the future (e.g., *The Jetsons*), fear ("finger on the button"), and perhaps especially convenience ("… with just the push of a button!"). Buttons populate electronic tools as well as toys from the pinball machine to the joystick. It is difficult to overestimate the historical significance of push button technology. DeRouchey calls it the "most influential yet least appreciated invention of the twentieth century." The interface for virtually every noteworthy technology today entails the pushing of at least one button. Some push button designs have become standards, like the circle and vertical line on a power button.

The 0–9 keypad is one of the most familiar and versatile designs of push button interface. The functionality of the telephone keypad surpassed the placement of calls, and the 0–9 keypad became the interface for ATMs, debit card readers, access panels and home security systems. Pushing buttons on a 0–9 keypad has become so commonplace that its historical significance is easily overlooked, especially now that we have entered the next era of interface. The touch screen era is upon us, although it's worth noting that iPhones continue to feature both a power button and a home button. More to the point, the keypad's layout is still recreated on smart phones, tablets and other touch screen technologies when numerical data is called for. Perhaps the most compelling evidence of the keypad's resiliency is its pixelated reproduction on touch screens.

Push button technology has no known inventor, but the telephone keypad does: Bell Telephone Laboratories. In March, 1964, Bell Labs issued a press release titled "The Touch-Tone® 'Dial.'" Five months after touch-tone's public debut, AT&T claimed the new method of placing calls was being welcomed across the country. "Typical comments" from "enthusiastic" (and anonymous) customers describe touch-tone as "very fast," "delightful," and "like magic." The new design had been tested for almost a decade, in order to maximize speed and minimize mistakes. Various sizes and shapes of keys were tested in sixteen different configurations of "rows, circles, triangles, crosses, even one very like the layout of the rotary dial," to determine their "speed, error rate and preference index." The "pushbutton action, or 'feel' came in for study" as well, because the engineers in Bell Labs knew that successfully implementing push button 'dialing' would hinge on "how our customers felt using the new telephone."[11] From the designers' perspective, what touch-tone calling would feel like was an exhaustive set of questions about touching a telephone. Inside Bell Labs the touch-tone problem concerned callers' fingers more than their feelings. AT&T's most famous advertising slogan, "Reach Out and Touch Someone," drew on physical contact to describe personal and emotional connections over the phone. During the holiday shopping season of 1988, AT&T celebrated touch-tone's silver anniversary by holding a three-day promotional event at its new flagship "Phone Center" store in Manhattan. Customers were invited to beam calls into deep space, and E.T. was on hand to "phone home" by manipulating new models with his famously illuminating index finger (which, not coincidentally, was too big for the holes of a rotary dial). In the film bearing his name, E.T. fixated on the telephone as a salve for his intergalactic homesickness, and his longing to "phone home" embodied the telephonic connection between feeling and feelings, between touching a telephone and reaching your people. No slogan or human spokesperson could rival E.T. when it came to demonstrating the capacity of touch-tone calling.

The touch-tone keypad is one of Bell Labs' most lasting accomplishments.[12] The field of Human Factors (popularly known as Ergonomics) also claims the telephone keypad as one of its founding achievements. Alphonse Chapanis, a "founder of ergonomics" according to his page-one obituary in the *New York Times*, was so intrigued by push-button dialing as a design challenge that he took a leave of absence from Johns Hopkins during the mid-fifties and went to work in Bell Labs.[13] Chapanis and Mary C. Lutz recruited three hundred test subjects and divided them evenly by gender and into three age brackets. They also divided their volunteers into self-identifying "naïve" and "sophisticated" users of technology. Once their subjects had been categorized, Lutz and Chapanis asked them to arrange the ten digits, zero through nine, into desirable configurations for a telephone keypad.[14] The six most popular arrangements were collected into further multiple-choice style tests, and Bell Labs continued to test other potential configurations for the keys. Articles published in *The Bell System Technical Journal* reviewed ongoing experiments that ultimately lead to the adoption of the keypad's winning design. The most comprehensive survey, published in 1960, began by announcing that "pushbutton signaling from the telephone set [is] within sight of economic feasibility," and set out to ask, "[f]rom the user's point of view, what are the desirable characteristics of pushbuttons?"[15] In most of Bell's tests, "right-reading 3 by 3 plus 1," now second nature across the wired world, was neither the fastest nor most popular design. It was one of the fastest, and one of the most popular, but a small percentage of subjects preferred two horizontal five-digit rows. An arrangement of buttons in a clockwise arc, like the finger holes in a rotary dial, was also more popular. Human Factors historian Henry Pertroski describes the ultimate design as "a judgment call and a compromise."[16]

During the mid-1960s, electronic calculators began replacing desktop adding machines alongside the first touch-tone telephones. The first companies to begin manufacturing digital calculators, however, "did not have the luxury of time that comes with a monopoly for conducting extensive human-factor tests."[18] (Legend has it that Bell Labs contacted Texas Instruments and other early manufacturers of calculators to ask if they would be interested in exchanging research. The calculator companies' response was, essentially, what research?) Calculators' keypads are numbered from bottom to top, and monopoly-funded labs are not the only explanation for the inverted top-to-bottom arrangement on telephones. The first Touch-Tone® phones were rolled out without letters (and without # or *), but Lutz and Chapanis tested A–Z alongside 0–9. Letters alongside numbers on telephones were nothing new, and while touch-tone phones were being introduced to the telephoning public, during the mid-1960s, the last exchange names were being retired. When the first touch-tone phones became available, dialers were still getting used to remembering each other's telephone numbers without the shortcut of words. All of the configurations that had tested best in Lutz's and Chapanis' formative study were arranged "in left-to-right order in horizontal rows starting with the top row" because people preferred the keys on their telephone to be arranged "in the order in which they normally read."[19] Numbers on the rotary dial featured letters, and their familiar presence helps explain the arrangement of the keys on the keypad.

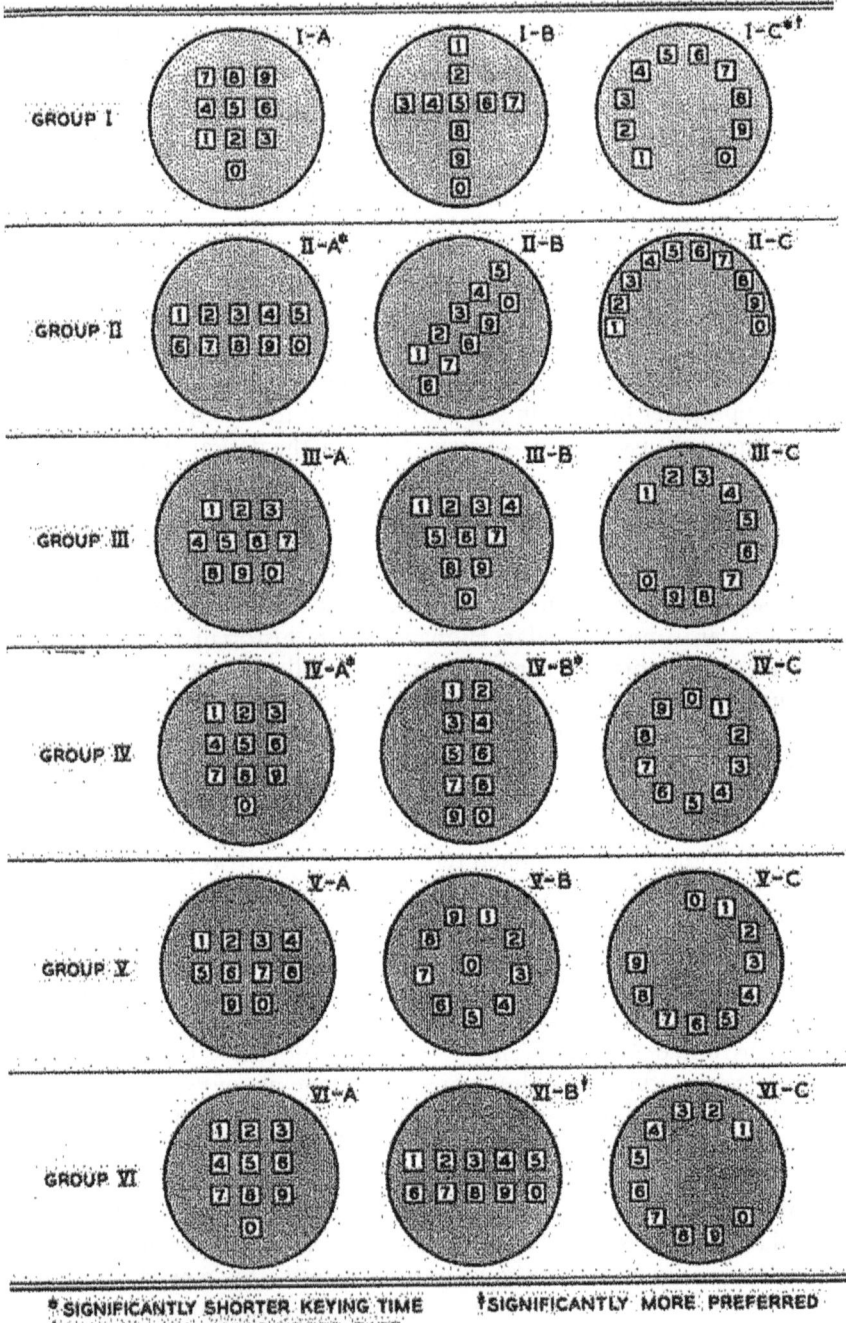

*Figure 3.2* Option IV-A: Neither the fastest nor most popular.[17]

Pushing buttons on touch-tone keypads sends DTMF (dual-tone multi-frequency) codes through telecommunications infrastructure, whereas turning a rotary dialing generates electronic pulses. What little touch-tone marketing there was touted the keypad as faster than dialing, but speed was a by-product of telephone connections being digitized. Whereas the dial had slashed labor costs for AT&T, touch-tone helped lower maintenance costs by eliminating nearly all of the "moving parts" involved in telephone technology. On the telephone itself, now only the twelve depressible keys themselves moved, with no internal mechanisms needed to generate rotary pulses. And on a vastly larger scale, shifting from electrical pulses to digital codes, as the medium for transmitting telephone numbers, enabled central telephone offices to complete the automation of calls. Digital infrastructure meant a computer could now activate and manage a subscriber's telephone number, rather than an electrician having to climb a telephone pole to activate or service the line manually. Electronic central offices were the true end game of automation for AT&T, and keypads rather than dials were the resulting interface for its subscribers.

Touch-tone calling required only a fraction of the training AT&T poured into the dial, and with little fanfare telephone users became accustomed to placing calls by pushing buttons. Over time consumers also came to accept and expect using the telephone keypad to access information and services. Apple's Siri and other voice-recognition software notwithstanding, companies providing automated customer service over the phone remain reluctant to abandon a keypad option. Push-button interface remains considerably less expensive, and the keypad "enjoys over-training in the U.S."[20] The telephone keypad's "over-training" stems from consumers' habitual use of it, not extensive promotion. As the previous chapter describes, Bell anticipated and encountered much consternation among their customers when they rolled out the rotary dial. The touch-tone commercial discussed at the beginning of this chapter demonstrates, conversely, how unconcerned AT&T marketers and executives were about keypads encountering similar backlash. The commercial's sleek visuals and sarcastic tone both exemplify the dominant advertising methods of the era, described by Thomas Frank as the "rise of hip consumerism," whereby advertisers opted to "level" with their target audiences rather than use hype and hyperbole to sucker them.[21] The tone of the commercial is a product of its era, while its hands-only demonstration of new telephone interface returned four decades later in the first series of commercials for the iPhone. The hands touching the iPhone's screen are again white, male and manicured, but the narrator's voice (and the soft music underneath) is soothing rather than sarcastic. Despite the different tones, the message in both commercials is the same: see how easy this is? Whereas the touch-tone commercial introduced the telephone keypad as nothing more than a faster way to place a call, the iPhone commercials celebrated everything made possible by swiping a touch screen. Whereas AT&T's touch-tone commercial implied (wrongly) that the keypad would simply enable faster calls, Apple tried to help viewers imagine everything you can do with the touch screen. "Reach Out and Touch Someone" has been replaced by: touch this screen to find anything you desire.

At the same time, the AT&T commercial was right – when it came to placing calls, the difference between a rotary dial and a touch-tone keypad was merely a matter of seconds. Cell phones have more dramatically revised dialing, with the naturalization of caller ID, number storage and shortcuts for favored numbers (all features introduced during the 1980s on touch-tone models). But the technological assemblage of a telephone call involves more than fingers and phone numbers. Technological assemblages entail more than bodies and machines; assemblages also affect and in turn are affected by customs, habits and attitudes. Thinking about technology as assemblage expands our purview beyond gadgets and bodies to focus attention on how *"practices, representations, experiences, and affects articulate to take a particular dynamic form."*[22] New technologies are always embedded within and interacting with other organizations, institutions, and infrastructures, and the touch-tone telephone was no exception. The keypad itself was invented and marketed by AT&T, and touch-tone phones were built for the monopoly by its manufacturing subsidiary, Western Electric. Subscribers leased phones from AT&T, and when they broke down a maintenance worker came and repaired (or replaced) it. When the AT&T monopoly was broken up, consumers began owning telephones outright. For the first time, subscribers of telephone service assumed responsibility for the purchase and maintenance of their hardware. After divestiture, as the next section describes, the home telephone was "reterritorialized" from leases and maintenance workers to purchases and personal responsibility.[23]

## Buying (into) Touch-Tone

In January 1982, AT&T entered into a Consent Degree to divest the company of its monopoly holdings. The U.S. Justice Department had filed an antitrust lawsuit against AT&T in 1974, on the grounds that "in the 1960s and early 1970s, it had thwarted competition in the marketing and manufacturing of phone equipment and the offering of services."[24] Bell controlled the markets for telephones and basic telephone service, and the steady spread of touch-tone over the same period was central to both claims in the lawsuit. Indeed, the Justice Department named touch-tone phones and the potential for new touch-tone services as justifications for breaking up the AT&T monopoly.

In the decade between the lawsuit and the Consent Decree, which went into effect on January 1, 1984, activists and regulators began reforming telephone policy. Both the costs of telephones and the means of possessing one were reformed, as customers were for the first time given the option of purchasing their own telephones, rather than leasing them from AT&T or a local subsidiary. In 1978, New York City Community Action for Legal

Services (NYC-CALS), a federally funded non-profit advocacy group, led a campaign to lower residential telephone rates, claiming that AT&T's customers were being overcharged both for their phones and for basic telephone services. Bell (both the inventor and the company named for him) initially imagined the telephone as a business tool, and only after decades of subscribers using phones to socialize did the company acquiesce and begin marketing the telephone as a social technology and their service as a way to "reach out and touch someone." A century after the telephone's invention, however, AT&T was still favoring its business clientele. NYC-CALS accused Bell of inflating costs for residential services, to the tune of $3 billion per year, in order to subsidize its business services. $120 million of the overcharges came from one service alone: touch-tone calling. AT&T was still charging its subscribers an extra "special feature" fee for touch-tone capability in 1978, priced up to three times the cost of rotary service, even though by then touch-tone was becoming commonplace among residential telephone subscribers in New York City and across the country.[25]

The lawsuit also claimed that Bell was bilking its residential customers with monthly rental charges for their phones, whether rotary or touch-tone. In New York at the time, it was "cheaper to buy a non-AT&T phone and throw it out every year than to pay AT&T's rental fee."[26] During almost a century of monopoly service, AT&T had leased phones to its subscribers, most of whom were unaware in 1978 that there were even "non-AT&T phones" out there for sale. Acting on pressure from groups like NYC-CALS, the New York State Public Utilities Commission "requested" that the New York Telephone Company, a Bell subsidiary, give its customers the option of buying their telephones, rather than paying monthly rental charges. New York Telephone was the first company to offer customers the option of buying the phones they had been leasing or purchasing new ones. In 1982, when the changes went into effect, monthly rental fees in New York City "range[d] from $3.05 for a standard rotary dial phone to $8.16 for a Trimline touch-tone phone," which at the time was Bell's most popular model.[27] AT&T set standard prices for the purchase of phones currently being rented. A standard rented rotary phone sold for $19.95, a standard touch-tone phone for $41.95, and a Trimline touch-tone model for $54.95. Now that activists and regulators were watching, AT&T also lowered its rental rates, with a standard rotary dial now leasing at $1.50 per month, a standard touch-tone for $2.85, and a Trimline touch-tone for $4.60.[28]

During the build-up to divestiture, local telephone companies and state agencies continued to enact piecemeal regulation, and touch-tone earned recognition as a basic mode of service, worthy of "universal access" in the language of federal telecom policy. The *Wall Street Journal* conceded that if basic telephone service "should be regulated (as most parties agree it should), then touch-tone, which is so intimately tied to it, should be also."[29] By 1986, only about sixty percent of all residential telephones were touch-tone models, but they dominated the markets of new telephone for sale.

Radio Shack (no stranger to push-button technology) led electronics retailers into the new market, and a wave of consumer advice columns ran in the nation's leading daily newspapers, with titles like "Phones: Buy or Rent?" (*Boston Globe*), "Consumer Saturday: Savings in Owning a Phone" and "The End of One-Stop Shopping for Telephones and Services" (*New York Times*), "New Bills Are Bad News to Phone Renters" (*Chicago Tribune*), and "Needlessly Paying Extra for Touch-Tone Dialing?" (*Wall Street Journal*). Consumers were typically encouraged to buy, on the assumption they could afford the one-time price, which would "pay for itself in 8 to 14 months." Newspapers also began listing the location and offerings of the new telephone stores, outlets with names like "Phone City" and the "Fone Booth" alongside chains such as Radio Shack.[30]

During the late 1980s and 90s scrutiny of monthly telephones bills intensified. AT&T's customers and consumer activists realized how much they had been paying for touch-tone phones, touch-tone service, and a host of new service options like call waiting and caller ID. Responding to pressure from consumer advocates and state regulatory commissions, AT&T and the "Baby Bells" began to itemize monthly bills, so that inflated charges could be identified and contested, beginning in California (a relatively progressive state regulator of telephony) in 1987. Throughout the nineties, other states eliminated monthly fees for touch-tone service, and some states even began distributing refunds for inflated fees. As late as 2003, telephone renters in Indiana were awarded a rebate for up to $80 in a settlement of a class-action lawsuit filed against AT&T, claiming that customers who rented their phones after January 1, 1986, had been charged inflated rates for them. Lawsuits have also been filed against phone companies who had been charging customers for touch-tone service even if they continue to use a rotary phone instead.[31]

However, after the AT&T monopoly was broken up, local subsidiaries began to raise their monthly rates again – no longer for touch-tone service, but for the continued rental of phones, which some subscribers still opted to do instead of purchasing a phone outright. In 1985, when people were first given the choice to buy, 40 million residential subscribers were renters. By 2006, 750,000 customers still rented their telephone, and two sisters in Ohio realized their grandmother was still paying AT&T $29.10 per month to rent her telephone. They calculated she had spent over $14,000 renting her rotary dial phone for the past 42 years. The sisters purchased a new touch-tone model for their grandmother, who told an AP reporter covering the story that she'd "like to have my rotary back. I like that better." Strogen was expressing a desire for one interface over another, rather than to leasing as opposed to owning. Her personal preference for an older technology over a new one was divorced from the commercial, market-based decisions about telephone use that consumers had to begin making during the break-up of AT&T. The newsworthiness of Ester Strogen, and her grandchildren's response to her leasing arrangement, demonstrates how thoroughly embedded telephones had become in commercial markets of pricing and regulation. During the touch-tone era, responsibility for negotiating these markets fell to consumers, just as the establishment of telephone connections had fallen to callers during the dial era.[32]

The Telecommunications Act of 1996 was a "massive overhaul [that] comprehensively amended" the foundational 1934 Telecommunication Act for the first time.[33] The 1996 Act is notorious for deregulating media markets and for massive giveaways of communications infrastructure to private companies, but it also included "universal service [a]s an explicit principle for the first time in law."[34] Elaborating on some general principles in the Preamble to the 1934 Act, the 1996 Act stipulates that the meaning of "universal service" will change, and that we cannot necessarily predict which services will become considered "basic." During the run-up to the Telecommunications Act of 1996, touch-tone was singled out for inclusion among basic services. Paul Zielinkski, President of the New York State Telephone Association, explained that not only had touch-tone become the standard means of placing calls, but also, as the rest of this chapter describes, touch-tone "is essential to allow people to obtain access to a host of services." Between divestiture and the 1996 Act, touch-tone calling became an aspect of standard telephone service for the majority of residential subscribers in the U.S.[35] By 1996, there was no doubt that touch-tone qualified, and the service was included in the F.C.C.'s definition of "universal service" without controversy.

Despite the savings that came with increased oversight of AT&T's pricing policies, not all consumers were happy about assuming ownership over their telephones. Pulitzer Prize-winning newspaper columnist Russell Baker, for one, described installing his new phone as a chore. In what he called the new "service economy … you, as the customer, are expected to double as the service department for the company that's taking your money."[36] Rather than promote repair and maintenance as frugal or environmentally friendly, let alone embrace a DIY ethos or celebrate his own technological self-sufficiency, Baker complained that the new service economy meant more work for him. He resented having to install his own phone instead of welcoming an AT&T employee into his home to do it for him. Activating a new phone was relatively simple and straightforward, but new to customers, and even distasteful to some like Baker. Unlike earlier aversion to dialing, described in the previous chapter, Baker was upset about taking on the blue-collar labor of installing his own new phone, after becoming accustomed to having someone do it for him. Whereas Senator Glass resisted being "transformed into a telephone operator without compensation," Baker objected to being conscripted to perform his own telephone maintenance. Computers may have taken over many of the routine tasks of managing telephone lines, but during the 1980s customers assumed responsibility for not only deciding how to possess their telephones but for installing them as well. Once telephone exchanges were fully digitized, a computer could activate a new telephone line, but someone still needed to plug in the actual phone.

Baker's resentment was intelligible to his readers, whether they shared it or not, in part because he tapped into lingering traditions of racialized and gendered entitlement among telephone users. Operators were trained to appeal to subscribers' sense of privilege, and echoes of Bell's "voice with a smile" slogan can be heard in Baker's complaint about installing his new phone. The title of his column is "Service with a Grimace," but now the frown is Baker's own, left alone to perform work he feels the company should be doing for him. Baker also picked up on the irony that, not only was he being left by AT&T to install his own phone, underserved and uncompensated, but he was also being charged an installation fee to do it. Baker concluded his column by bemoaning the fact that in the new service economy, "we have ended up with telephone companies that charge us for doing their work." Baker's "service economy" is more accurately a self-service economy, and while his complaint reeks of elitism, he also put his finger on the costs and responsibilities for consumers that were beginning to accrue to digital telephone technologies and the new markets surrounding them. Like dialing before it, the break-up of AT&T's monopoly changed everyday telephone use in ways that meant new costs and responsibilities as well as tasks for telephone users. Baker highlighted some of the manual labor being reassigned to consumers, and he exposed AT&T's attempt to generate revenue as well as saving by conscripting its customers. While the decision to buy or lease a telephone increased customers' autonomy over their telephony – a key argument for ending the monopoly – Baker also points out that the new market of telephones for sale introduced new responsibilities as well as choices.

The touch-tone keypad was invented in Bell Labs, while the standardization of touch-tone calling in the U.S. resulted from AT&T's monopoly divestment. The technological assemblages featuring touch-tone extend beyond fingers pushing buttons and include the largest institutions of business and government. The interface transition from rotary dials to touch-tone keypads was catalyzed by "the largest corporate shakeup in world history," and in its aftermath consumers *en masse* assumed ownership of their telephones for the first time.[37] Telephone service was a century-old industry when ownership was delegated to subscribers, and in hindsight this assumption of responsibility for one's own phone helped prepare the dialing public for cell phones and contracts with cellular service providers.

## The Dialing Gap

After exercising monopoly control over virtually every aspect of telephony in the U.S. for a century, AT&T agreed in the Consent Decree to part with over twenty of its regional subsidiaries and all of its manufacturing divisions. However, AT&T would keep its most lucrative service, long-distance calling, and would also be allowed to compete in other markets of the expanding telecommunications industry. (AT&T also retained Bell Labs, after the Department of Defense worried publicly that divesting one of the most prominent telecommunications labs in the U.S. would pose national security risks.) Observers of the telephone industry predicted that divestiture would "help AT&T, hurt consumers and have mixed results for AT&T's numerous competitors." The *New York Times*, in one of two front-page stories covering the settlement, on January 9, 1982, quoted a telecom analyst who insisted, "[w]hat they've done is taken the most capital intensive, politically intensive, labor intensive part of their business and given it to someone else." In having its monopoly divested, AT&T managed to dump its least profitable service, local service; keep its most profitable one, long-distance; and get federal permission to compete in other markets that it had been barred from entering. In 1989, for example, AT&T was cleared to begin offering email and other "electronic publishing" services.[38]

The divestiture of AT&T went into effect on January 1, 1984, and during the lead up telephone customers took notice of their telephone bills like never before. The newspaper columns about buying or renting phones, mentioned earlier, had been only the beginning. As the break-up approached, newspapers analyzed it for readers in terms of what it would do to their phone bills. The *New York Times* announced the "end of one-stop shopping for telephones and services," meaning the entirety of telephony would no longer be provided by one company and charged on one monthly bill. By the time of the divestiture, consumer activists and state regulators had already forced most telephone companies to itemize their bills, so customers could see each and every thing AT&T was charging them for, and how much each piece of telephone equipment and service cost. But now several companies would provide elements of basic telephone service, and the bills would come from different companies. On the eve of divestiture, in December 1983, the *Boston Globe* assured its readers that "the bottom line won't be much different," but also warned them on page one that their next round of telephone bills was "going to be a doozy." If you still rented a phone, then a bill for it would still come from Western Electric or another manufacturer, but now the charge for local service and wire connections would come from one of Bell's divested subsidiaries. Furthermore, now a separate company would provide long-distance service.[39]

The consumer market for long-distance telephone service opened in March 1980, when MCI Telecommunications, Corp. began offering residential long-distance calls. MCI had for years been offering national and transcontinental long-distance service, called "Execunet," to businesses and corporations in the largest few dozen U.S. cities, at rates far lower than AT&T's. MCI was able to undersell their monopoly competitor because they used microwave stations, spaced twenty miles apart, to transmit the data of telephone conversations. The microwave stations could beam a call for much less than what it cost to send calls through telephone wires, but the microwave stations could neither initiate nor complete a call. The first and last relay still had to travel over local telephone infrastructure, so in order to receive and connect calls, MCI needed to purchase local telephone service from AT&T, which controlled over 99% of local service. In 1975, the F.C.C. had ruled that AT&T was no longer required to provide the "final mile" of local connections for MCI, because the telecom upstart was building Execunet into a long-distance telephone network essentially no different than Bell's system. MCI appealed the F.C.C.'s decision and won; AT&T and the F.C.C. appealed, repeatedly, and lost all the way to the Supreme Court, who in January 1978 refused to hear an appeal to a lower ruling, which required AT&T to continue providing its local telephone service to MCI. 1978 was also the year Bell began facing lawsuits about overcharging residential customers in order to subsidize their business services, which were much more intricate and extensive, and therefore more lucrative even at cut rates.[40]

MCI crept into the long-distance market with Execunet. Sinking telephone networks and hardware into business services was profitable because of the sheer volume of calls made during business hours. However, this left MCI's Execunet infrastructure "underutilized at night," so when the Supreme Court cleared the way, MCI went residential. AT&T had already trained its subscribers to schedule their long-distance calls during evenings and weekends, by offering reduced rates when the monopoly's networks did not have so many business calls to handle (a pattern repeated initially by cell phone carriers). It would not be easy to lure subscribers away from a century-old monopoly, and so to attract new customers MCI went on the offensive and promoted its lower prices. The company's first slogan for its residential service was "you're not talking too much, you're paying too much." For a subsequent campaign, when the long-distance market had been saturated and marketing grew more aggressive, MCI recruited Joan Rivers to attack AT&T more directly. Opening with her trademark, "can we talk?" Rivers mocked the monopoly with its most famous slogan: AT&T was "reaching out in the same way they always have – for your wallet." MCI blanketed airwaves and billboards with promises that you could save 25–50% on your monthly bill by leaving AT&T, and most observers agreed with the figures (even AT&T spokespeople when pressed).

The only catch was that placing a long-distance call with one of the new carriers required dialing twenty-one or twenty-two digits, as opposed to ten with AT&T. MCI customers would not only have to dial the telephone number of the person they were trying to reach, but also a series of access codes for MCI's infrastructure.[41] MCI filed federal lawsuits to close what their lawyers and spokespeople referred to as "the dialing gap" between AT&T's established "1+" system of area codes and their own more cumbersome systems. At the time, this phrase was an obvious play on the "missile gap," that popular bogey of Cold War America, but it resonated for another reason. During the telecom battles over "the dialing gap," a long-distance call was still the only reason most people in the U.S. would enter more information into their phone than the seven digits required to place a local call. MCI and its competitors for Bell's long-distance business recognized the added dialing requirement as their single, biggest obstacle to a full share of the long-distance market.

The opening of the long-distance market hastened the demise of the dial. In order to access one of the new long-distance carriers, customers would not only have to dial more numbers, they would have to use a touch-tone phone. Customers like Ester Strogen would be stuck with AT&T, while "customers who choose to use MCI or Sprint [had] to use a touch-tone phone and punch in lots of extra numbers." MCI's system of microwave transmitters and the computerized network connecting them were designed to read the DTMF codes produced by touch-tone phones, but not the electronic pulses generated by turning a rotary dial. Consumers wanting to say goodbye to AT&T's overpriced "LongLines" service and try one of the upstarts would need or purchase (or lease) a touch-tone phone. In two years MCI attracted 850,000 customers to its residential service, but there were still over sixty-five million rotary phones in U.S. homes. Telephone manufacturers, racing to fill the newly expanding market for telephone hardware, knew that they were fighting for the same customers who were also shopping for long-distance carriers offering touch-tone savings. Once MCI began offering residential service, a cottage industry sprang up for devices that converted dial pulses into DTMF tones, "allow[ing] both rapid dialing and access to many of the new phone systems." Not only did some companies begin selling phones that could translate rotary pulses into touch-tone codes, but at least one company tried to pass off rotary technology with touch-tone interface. In 1986, the F.T.C. settled out of court with Cosmo Communications Corp., after charging that "although the company's push-button phones look like touch-tone telephones, many of them operate like rotary dial phones and therefore can't connect to computerized banking services and long-distance services."[42] The difference between the dial and the keypad as interface experiences was negligible compared to the latter's capacity to access digitized services.

The rotary dial was a problem for companies like Sprint and MCI, a technological barrier to substantial segments of the long-distance market. Another problem was that too many telephone customers refused to choose a long-distance carrier. When AT&T divested, its subscribers were automatically switched to whichever "Baby Bell" was taking over local service in the area. Telephone subscribers hardly noticed the shift from one local telephone company to basically the same company with a different name, but the end of AT&T's monopoly over long distance became a more complicated matter. It is less accurate to say that opening the long-distance market meant that AT&T's subscribers *could* choose a new carrier, than it is to say they *had to* choose. Four months after the Consent Decree went into effect, telephone companies and the F.C.C. both admitted that "[t]he latest phase of telephone deregulation is not working as well as expected ... because most customers are ignoring requests to choose between the American Telephone and Telegraph Company and nearly a dozen competitors for their long-distance service." According to regulators and officials, four months after divestiture, roughly half of telephone subscribers nationally had failed to pick a new long-distance carrier.[43]

One intention of the Consent Decree was to open the long-distance market, so subscribers would only be able to stay with AT&T if they actively choose to do so. Local telephone companies sent their subscribers "Easy Access Shoppers Guides" and mock ballots with a complete list of long-distance options. The California Public Utilities Commission, for one, mandated that these consumer resources be sent at least 90 days before a local exchange's changeover. MCI's executives and those of other new long-distance carriers complained that AT&T and the "Baby Bells" were dragging their feet on their federally mandated "equal access" upgrades, and some upstarts sued over a plan to automatically resign subscribers with AT&T if they failed to choose in a timely fashion. Finally, the F.C.C. set May 31, 1985, as the deadline for subscribers to choose. The Courts and the Justice Department had made it clear to the F.C.C. that AT&T would not have any unfair advantages in the newly competitive long-distance market, so those callers who had failed to choose by the deadline would be randomly assigned a long-distance carrier by their local telephone company.[44]

Divesting AT&T's monopoly did not simply deregulate the telephone industry; it created a new commercial landscape in which law and policies would have to be created to police consumers' telephone practices. Consumers' found themselves with new decisions to make about their telephones, such as whether to buy or rent and which company to use for long-distance service. The choice between purchasing a rotary or touch-tone phone was essentially made for them, thanks to the growing obsolescence of rotary technology, but the new market in telephones for sale brought added responsibility as well as costs. Subscribers began maintaining their own telephones, *ala* Russell Baker, and managing their own telephone use through a new set of atomized and itemized bills. The break-up of AT&T's monopoly and the "end of one-stop shopping for telephones and services" meant that consumers needed to begin performing the cost-benefit calculations that comprise comparison-shopping in a competitive market. The combination of digitization, on the one hand, and on the other hand markets in telephone service becoming competitive, meant that the new forms of consumer labor accumulating around everyday telephony were not strictly a series of mental and manual tasks, as had been the case with the dial. Instead, during the touch-tone era, deregulation of telecommunications entailed a more diffuse and elaborate transfer of costs and responsibilities. The new market in long-distance service was one of the first instances of consumers and commentators taking note of these new responsibilities and recognizing them as obligations as well as options.

Many telephone customers needed to be prodded repeatedly before selecting a long-distance company, while others appreciated the savings possible in a newly competitive market. One *New York Times* reader was moved to write a letter to the editor in response to an article about the costs of phone calls placed from hotel rooms. Published in October, 1982, the letter was penned by Bernard L. Albert, M.D. of Scarsdale, NY, who explained that the new market of "private long-distance services" provided an "answer to this problem."[45] Unlike Russell Baker, Dr. Albert welcomed the new array of choices and responsibilities facing him in a digitized and deregulated telephone industry. His letter celebrated rather than criticized the multiplying relationships between telephone users and service providers. Hotels at the time inflated their telephone rates to generate revenue from business travelers, a captive market before mobile phones who tended to accept the rates when their employers were footing the bill. Dr. Albert was replying to an article aimed at tourists and leisure travelers; but even on holiday, the good doctor knew to bring a businesslike savvy to his long-distance decision-making, and he was happy to dial extra digits in exchange for savings and convenience. Dr. Albert's letter also conveyed an appreciation for touch-tone as the technological upgrade enabling his long-distance savings. He mentioned the feature by name, capitalizing it, and even alerted readers to the "tone generators" that could be purchased, if necessary, as a small investment toward long-distance savings. Dr. Albert identified himself as a customer of Sprint in particular, and he praised his carrier as much for the convenience it enabled as the savings. Dr. Albert expressed appreciation for the new long-distance service option because it allowed him to avoid "the often exorbitant surcharges and high telephone company operator assistance rates" that attached to calls from hotels, and he also celebrated no longer needing to carry cash, specifically the "large number of coins" needed to feed a pay phone. He relished the enhanced access and flexibility that came with the capability of placing a longdistance call from any phone in the country and much of the world. He essentially presents the convenience of long-distance access codes as a form of savings. Switching to Sprint for your long-distance calls will save you time as well as money, Dr. Albert promised.

One can imagine how gratified Dr. Albert felt when, a few years later, AT&T began discounting self-service calls ("direct-dialed long distance rates") and raising rates for operator-assisted calls, including all international calls. This pricing strategy by AT&T marked the first large-scale attempt of the touch-tone era to tax customers requesting personal assistance, in order to subsidize those ready, willing and able to serve themselves. At the dawn of the touch-tone era, Russell Baker recognized installing his own telephone as an imposition, a burden, and one that AT&T had the audacity to charge him for shouldering. Conversely, Dr. Albert's letter to the editor heralds a new day, and he represents the ideal new neoliberal telephone user, grateful to a company for allowing him to take more control over his long-distance telephone calls, rather than having to subject himself to the layers of costly intermediation provided by telephone companies (and hotels). Dr. Albert embraced the dialing gap and conceived of it as an exchange for savings made possible through market competition.

All consumers were faced with new decisions about their telephone use, and unlike Dr. Albert, many callers seemed reluctant to take them on. Placing a long-distance call with one of the new companies meant dialing (and remembering or storing) twice as many numbers as it had with AT&T's patented "1+" system of area codes. The telephone advice columns explained that, until your local changeover occurred, you would have to keep dialing "a finger-daunting string of digits" to place your long-distance calls with any company other than AT&T.[46] Some customers, like Dr. Albert, were happy to do so in order to save time and money. Meanwhile millions of AT&T subscribers refused to choose a long-distance carrier after the break-up. (Other customers, such as Ester Strogen, never wanted to upgrade to touch-tone technology or enter its new markets in the first place.) Once federally mandated "equal access" had been achieved from coast to coast, all long-distance carriers could offer 1+ ten-digit access system. Consumers could then choose a "primary carrier" and receive monthly bills from them. But, in the spirit of deregulation, it was also decided that callers should be able to choose a long-distance service provider for each call. All long-distance companies were granted a three-digit prefix for customers to use as an "access code" to the particular companies' long-distance service. AT&T became 288 (A-T-T in alphanumeric telephone code); MCI became 222, and SPRINT (by then GTE-Sprint) became 777. Callers did not take offense at this "creeping numeralism," as the anti-digit dialers from the previous chapter may have, nor were there any "Stop the Overlay" style campaigns to fight these area code giveaways. The three-digit prefix codes had already proven to be acceptable, to callers and regulators alike, as a form of telephonic categorization. Area codes had been in use for several decades when the long-distance market was rationalized. The long-distance "access codes" applied prefixes for the first time to companies rather than geographic areas. In the wake of AT&T's divestiture, alongside the long-distance start-ups, other companies began offering new automated telephone services (ATS), fellow pioneers settling the wild west of a digitized and deregulated telecommunications industry. Initially, as the final section of this chapter describes, the ATS would be regulated by through the use of commercial prefix coding. Ultimately, a new regulatory distinction would emerge between dialing a telephone number and "dialing" additional data, such as a long-distance code or, more significantly in the long run, a credit card number.

## Press 1 for Payment

In a 1995 episode of *Seinfeld*, Jerry's neighbor Kramer discovers that his home telephone number is one digit removed from the number for Moviefone, an ATS designed to help patrons locate theaters and show times.[47] Moviefone was new at the time and only available in New York and Los Angeles, but in 1999 AOL purchased it for over $400 million dollars and soon partnered with Movietickets.com to become a transactional as well as informational

ATS.[48] Now callers could key in credit card information to purchase tickets as well as navigate menu options to find showings. But Moviefone was still an informational ATS of the old guard when the *Seinfeld* episode aired. Inundated with wrong numbers, Kramer decides it is more bother to redirect callers than to provide the information they seek. Inevitably Jerry's friend George misdials and reaches Kramer, who mimics Mr. Moviefone's voice: "using your touch-tone keypad, enter the first three letters of the movie you'd like to see." The DTMF tones of touch-tone keypads are decipherable to fewer human ears than Morse's code ever was, and when George complies, of course Kramer is unable to decipher the tones, so he takes a guess and inquires/instructs: "if this is correct, press one." When George does nothing, Kramer guesses again, and again, until finally he pleads, "why don't you just tell me the name of the movie you'd like to see?," a plea that anticipated voice-recognition software, which many ATS like Moviefone began offering as an alternative to keypad data entry.

When dialing emerged as the interface for automated connections, telephone numbers were already a systematic form of individual identification. From the perspective of hindsight, the telephone number was in many ways more pivotal than the dial itself. Telephone numbers preceded the dial and helped prepare callers for dialing. When AT&T automated its local service, callers no longer established a connection by speaking a sequence of letters and numbers to another human being; instead, callers began using their index fingers to mechanically communicate the sequences. Bell introduced telephone numbers in part to "distance" its subscribers from the personal attention of operator service to which they had grown accustomed. Identifying one another by numbers instead of names made it less of a leap for callers to then begin dialing those numbers instead of saying them to an operator. Telephone numbers were introduced during the 1880s and 1890s, making them older even than the driver's license, which dates in the U.S. to the 1910s. After the postal address (and alongside the passport), the telephone number was one of the first systemized forms of identification in the U.S. Telephone numbers were also an industrial-era precursor to the more recent proliferation of personalized data and information – credit card numbers, PINs, access codes, and passwords – transmitted via digital interface.

After some initial resistance, callers by and large accepted telephone numbers to the point that, ultimately, many residents embraced exchange names and area codes as signifiers for hometowns and neighborhoods. The previous chapter introduced the Anti Digit Dialing League, a 1960s activist group who formed to fight the retirement of exchange-name prefixes in favor of "all-digit dialing." The ADDL adopted the phrase "creeping numeralism" as a rallying cry. Today their distinction between names and numbers feels outmoded, overemphasized, and perhaps even misguided. With txt msg, telephony became fully alphanumeric, and thanks to smart phones it is now audio-visual as well. But in the ADDL's defense of telephone exchange names as a source of identity as well as identification, we can glimpse the future they were fighting – our present – within which alphanumeric identifications continue to pile up. What would ADDL members, for instance, make of email addresses and Twitter handles? On keypads and keyboards, the pound sign now does double-duty as the hash tag, prefix punctuation for identifications chosen by consumers rather than assigned by service providers. The historical trajectory from the pound sign to the hash tag helps explain the commercial and cultural development of what Mark Andrejevic calls "digital enclosure," whereby information rather than land is the resource expropriated by private interests for commercial gain.[49] It was during the touch-tone era that callers began to enter information other than a telephone number into their phones. Long-distance access codes were one of the first instances of touch-tone data entry, although technically the codes were an extension of telephone numbers, in that callers keyed them to access a particular company's infrastructure *en route* to another telephone. The first touch-tone data entry beyond a telephone call was the keying of selections among ATS menu options. ATS were the first assemblages in which callers used the letters and numbers on a telephone keypad, as well as * and #, to communicate anything other than a telephone number.

The Consent Decree that divested the monopoly included "information services" among the telephone businesses now open to competition.[50] The first automated information lines to be made available to the dialing public included weather reports, stock tickers, and sports scores. Before the Internet (and cable TV) became a clearinghouse of outlets for up-to-the-minute information, federal agencies, legislators, and judges clashed over how to regulate citizens' access to automated information over the phone. The federal attention paid to ATS took place amidst an unprecedented wave of deregulation throughout the telecommunication industries, leading up to and in the wake of the AT&T break-up. The commercial relationships between providers of automated services and the companies who owned the telephone lines delivering the information came in for oversight during this period. Issues involving new uses of telephone infrastructure were affected by attempts to police the illicit content of some service providers. Initial attempts to regulate digitized telephone content took place in the shadow of television and other mass media, and they also occurred shortly before personal computers and cell phones became popular access points to a wealth of online information and services. The regulation of touch-tone telephones, as the interface for ATS, helped set the stage for regulatory battles over access to and ownership of digital content online.

In 1989, less than a year after E.T. had helped AT&T celebrate touch-tone's silver anniversary, newly elected President George H. W. Bush appointed Alfred Sikes as Chair of the Federal Communications Commission. Sikes took office vowing to "have touch-tone phones installed throughout the agency by the end of the year to improve service to licensees, parties [and the] public." In 1989 the U.S. government was lagging behind other federal governments and state telephone agencies in capitalizing on the new technology. In 1984, the Nippon Telegraph and Telephone Public Corporation, Japan's "semi-governmental telecommunications monopoly," purchased 60,000 touch-tone phones from a Tennessee supplier, the first NTT purchase of telephones manufactured in the U.S. "Automated information services" quickly became a growth industry in Japan, and soon the national telephone authorities in both England and France began collaborating with "independent information providers to generate commercial revenue from selling ... information over the telephone." France's state telecommunications authority, the DGT, introduced its "Audiophone services" in 1983. Two years later, over 1,300 automated information services throughout France, such as a "speaking clock" and news briefings, had been called over one hundred million times throughout France, generating Ff 115 million ($14 million) for the DGT in 1985 alone. In addition to these public services, the DGT also began working with private companies to provide "premium rate services," for which a caller would pay a fee on top of the standard cost of a telephone call. Across the channel, British Telecommunications (BT) did not take a cut of the profits flowing from audiotext services, but planned on increasing its revenue by expanding uses for their basic telephone network. BT also began teaming with VoiceCom Systems, Inc., a private company in the U.S., to provide international messaging services between the U.K. and Europe.[51]

The U.S. government was a latecomer to ATS provision, but federal pressure to police ATS content amidst the deregulation of the telecommunication industry continued to build during Alfred Sikes' tenure at the F.C.C. from 1989 to 1992. Once companies like VoiceCom began using touch-tone and DTMF signals to sell new services (or old services newly over the phone), all three branches of federal government in the U.S. struggled to determine who had jurisdiction over which aspects of the nation's automated business and pleasure. Elected officials shared the F.C.C.'s goal of keeping the ATS industry unfettered, but the U.S. government never worked with those companies to generate public revenue or services, such as their British and French counterparts did. Members of Congress balanced their traditional role as business boosters with laws to protect voters and, especially, children from the risks and dangers of audiotext. And while the F.C.C. under Sikes did not want to interfere with companies trying to provide automated telephone services, Congress and the Supreme Court clashed during Sikes' tenure over how to regulate the content now available in the new ATS industry.

It was not long after automated information about stocks, sports and weather became available over the telephone that "purveyors of phone sex [became] the leading innovators in the marketing and technology of pay-per-call services" in the U.S. Some ATS lawsuits were asexual (such as one against a TV station in Seattle for broadcasting DTMF tones so children without a touch-tone phone, or toddlers who had yet to master dialing, could connect to a Talk-to-Santa ATS by holding a phone up to a television set), but most cases involved adult-oriented content. Soon "angry parents [were] on the warpath, suing telephone smut purveyors from California to Cleveland." "Dial-a-porn" was a predecessor to live two-way phone sex lines that quickly became the most profitable segment of the ATS industry. During the culture wars of the late-1980s, pornographic ATS were subject to state and federal lawsuits as well as new legislation. To access dial-a-porn consumers performed the same tasks and pressed the same buttons as they did to check the weather, but federal policy and legislation treated sex lines differently from other ATS. Dial-a-porn legislation struggled to accomplish two goals at once: to police audiotext content while keeping unfettered the burgeoning business in ATS. In doing so, aspects of the regulatory framework were established for accessing content online as well as for providing the personal information of Andrejevic's digital enclosure.

The first lawsuits were filed during 1987 and financed by Citizens for Decency through Law, an anti-obscenity group based in Scottsdale, Arizona, which later became the first state to pass its own legislation policing dial-a-porn. Jessie Helms (R-NC) famously proposed legislation to ban pornography in virtually any format within federal jurisdiction. Culture wars over arts funding took center stage, but Helms' bill also contained "provisions affecting broadcast, cable, programming, publishing, satellite, telephone, [and] videocassette industries," and it amended the 1934 Communications Act to outlaw dial-a-porn in particular, because of the sexualized threats and opportunities presented to children by new adult-oriented ATS. Helms' proposal passed the Senate with a 97–0 vote, soon it passed the House, and dial-a-porn was illegal. Selling indecent prerecorded messages over the phone was now punishable by up to two years in jail and fines of up to $500,000. In November 1988, the F.C.C. issued its first fine for trafficking in dial-a-porn, although the penalties were probably "more valuable to [the F.C.C.] as symbol than as an actual enforcement move," since the fined company had already gone out of business.[52] In January 1992, Telesphere Communications Inc., the largest provider of dial-a-porn, filed for bankruptcy, signaling a new day for the ATS industry. Peter Brennan, director of the audiotext firm Tele-Publishing Inc., claimed that "the industry is actually shifting away from such entertainment lines to business-to-business and business-to-consumer services." The *New York Times* announced that the ATS industry was graduating from "the racy [to] the respectable."[53]

Dial-a-porn lines focused legislative attention on the fledgling ATS industry, and local telephone companies responded by using prefix codes to classify different types of lines. Prefix codes meant the local phone companies could charge service providers Information Delivery Service tariffs, variously priced according to the genre of service being accessed by callers via their local infrastructure. For instance, in order to clearly and consistently demarcate "indecent" ATS services, in December, 1991 the New York State Public Utilities Commission ordered all dial-a-porn services in New York City to identity themselves with a 970 prefix. Four months later, a federal judge ruled the New York [City] Telephone Company could disconnect five dial-a-porn services without violating free speech rights because their crime had nothing to do with content. Local telephone exchanges were responsible for delivering pornographic messages into home telephones, alongside the long-distance calls beamed across the country by Sprint and MCI, and the judge allowed the local exchanges to cut off the five companies because they had misidentified themselves with 900 prefix codes. The NYS PSC used the 970-prefix to segregate dial-a-porn in an ATS ghetto. The restriction, however, applied only to prerecorded audiotext messages; the 900 prefix was still legal for the new, vastly more popular and lucrative live-talk "phone sex" lines. Most interactive phone sex service providers used 900 numbers, an expansive and upwardly mobile telephone prefix. Unlike dial-a-porn and other 970-lines, which had to piggyback onto callers' monthly phone bills, 900-line services were legally allowed to ask callers for credit card numbers to which the services could be billed directly.

The Senate continued to weather a "storm of consumer protest" about ATS, but Ernest Hollings (D-SC), chair of the committee overseeing telecommunications policy, declared that during 1991 "the No. 1 consumer complaint in America" was ATS scams rather than illicit content. Federal policing of ATS fraud was handled by the F.C.C. and by the Federal Trade Commission, which also began filing claims for "consumer redress" against fraudulent ATS companies. Laws against ATS fraud were passed at the state level, but state legislators ignored ATS content, focusing instead on protecting voters from receiving unwanted calls placed by telemarketing "auto-dialers."[54] The 900-telephone prefix, which callers began using to interact and transact with service providers, was an important and ingenious technology of consumer labor in its own right. During the late 1980s and early 1990s, consumers began calling 900-numbers and entering their credit card numbers into touch-tone keypads. Telephones had been used to discuss, arrange and agree to transactions since the days of operators, but the touch-tone keypad was the first everyday interface capable of facilitating transactions at a distance.

By the time the 1996 Telecommunications Act passed, a year after the *Seinfeld* Moviefone episode aired, the focus on protecting children from harmful or indecent content had shifted from ATS to the internet. The 1996 Act was challenged in federal court, and appeals were heard all the way to the Supreme Court. Some touch-tone rulings had direct implications for internet traffic. In one case, an adult-oriented ATS firm complained that their free speech rights were violated when regional telephone companies stopped completing their calls. The Court's ruling used audiotext services as a bellwether for looming issues about internet policy and the policing of access to content online. The landscape of media policy shifted when the Court determined that logging on was more like dialing a telephone than turning on a radio or TV. In describing telephonic access to sexual content, the Court cited a lower District Court ruling that "communications over the internet do not 'invade' an individual's home or appear on a computer screen unbidden." Neither the Internet nor the telephone, the Court maintained, are as "invasive" as either of the old broadcast media.[55]

Disclaimer guidelines from the ATS industry were applied to the internet, and the Court found "read before you enter" warnings to be sufficient disclosure of the content therein, as well as proper protection of minors from pornography. "[A]lmost all sexually explicit images are preceded by warnings as to the content," the Court explained, and it quoted testimony claiming that the "'odds are slim' that a user would come across a sexually explicit site by accident." The Court established a regulatory distinction for the ATS industry between the active steps required to dial a phone and the passive reception of radio or TV. "Placing a call," Judge Stevens wrote, echoing the 1996 Act "is not the same thing as turning on a radio and being taken by surprise." (In light of news feeds and other aggregators of online content, the distinction is *passé*.) The F.C.C., the Supreme Court, and other touch-tone regulators found in the internet reflections of older media, and they conceived of and ruled over the new services in relation to the old. Online content was deemed more similar to audiotext messages than to mass media programming delivered via radio and TV. The Supreme Court first ruled on the constitutionality of internet porn by comparing it to dial-a-porn and contrasting both to the "passive" reception of broadcast TV and radio programming.[56]

In its early internet rulings, the Supreme Court treated dialing a phone number as activity, if not work, in excess of turning on a radio or TV to receive broadcast information and entertainment. The Court policed dial-a-porn lines by separating the ATS industry into two sectors: a "dial-a" sector, featuring prerecorded audiotext messaging such as news updates, stock prices, and sports scores; and a "dial-it" sector on the other, in which customers called ATS lines not merely to listen, but to talk as well, including "phone sex" lines. Dialing was enough to access dial-a audiotext content, whereas dial-it ATS required callers to key in a credit card number in order to access content. The "a" in "dial-a" was a prerecorded message; the "it" in "dial-it" was not the content heard, but any content spoken by the caller or any data entered into a keypad. The Supreme Court carved out dial-a-porn for oversight by creating a new category to legally separate it from the "dial-it" services offering live-voice interaction and transactional data entry. By 1996, dial-a-porn accounted for no more than that 4% of the ATS industry, yet this technologically outmoded service (interactive "phone sex" had taken over) came in for the lion's share of regulation in the new Act.

The Helms Amendment and the subsequent legal battles were about the regulation and policing of information and entertainment content accessed over the phone. Ultimately, the Court adjudicated the ATS industry by drawing a distinction between dialing and digital interface. The legal recognition of keypad data entry beyond dialing telephone numbers, specifically to enter credit card numbers, catalyzed the keypad's emergence as an everyday interface for digitized transactions. The first money machine (beyond the telephone) to feature a keypad was the ATM, subject of the next and final chapter.

# Notes

1. Commercial retrieved from YouTube (and since removed). Screenshot from recording in possession of author.
2. Thanks Adam Rottinghaus for bringing Louis' joke to my attention.
3. Dan Schiller, *Digital Capitalism: Networking the Global Market System* (MIT, 1999): 24.
4. Part two of Venus Green, *Race on the Line* is titled "The Dial Era, 1920–1960."
5. D. Schiller, *Digital Capitalism*: 7.
6. For leading analyses of the break-up, see Harry M. Shooshan III, ed., *Disconnecting Bell: The Impact of the AT&T Divestiture* (Pergamon Press, 1984); Sam Simon, *After Divestiture: What the AT&T Settlement Means for Business and Residential Telephone Service* (Knowledge Industry Publications, 1985); Steve Coll, *The Deal of the Century: The Break-Up of AT&T* (Antheneum Press, 1986); Fred Henck, *The Slippery Slope: The Long Road to the Break-Up of AT&T* (Grenwood Press, 1988); Alan Stone, *Wrong Number: The Break-Up of AT&T* (Basic Books, 1989); and Barry G. Cole, ed, *After the Break-up: Assessing the New Post-AT&T Divestiture Era* (Columbia UP, 1991).
7. See Patricia Aufderheide, *Communication Policy and the Public Interest: The Telecommunications Act of 1996* (Guilford Press, 1999).
8. Lisa Nakamura, "What Steven Wants: Gestural Computing, Digital Manual Labor, and the Boom! Moment," *in media res: a media commons project*, March 11, 2008, accessed October 22, 2011. http://mediacommons.futureofthebook.org/imr/2008/03/11/what-steven-wants-gestural-computing-digital-manual-labor-and-boom-moment.
9. Ibid.
10. Vincent Mosco, *The Digital Sublime* (MIT Press, 2004), 7.
11. "The Touch-Tone® 'Dial,'" Bell Telephone Laboratories press release, March 1964.
12. AT&T consolidated its operation and engineering, and research and development departments into Bell Telephone Laboratories in 1924. In the previous chapter, I used "AT&T" and "Bell" interchangeably, although by the time touch-tone was introduced in the mid-1960s, "Bell" had largely given way to "AT&T" as the popular name of the telephone monopoly, and in this chapter I will follow suit.
13. S. Lavietes, "Alphonse Chapanis dies at 85; was a founder of ergonomics," *New York Times*, October, 15, 2002, A1.
14. Mary C. Lutz and Alphonse Chapanis, "Expected locations of digits and letters on ten-button keysets," *Journal of Applied Psychology* 39 (1955): 314–17.
15. R. L. Deninger, "Human Factors Engineering Studies of the Design and Use of Pushbutton Telephone Sets," *Bell Technical Journal* 39 (1960): 995–1012.
16. Henry Petroski, "Human Factors," *American Scientist*, 88 (2000): 307.
17. Deninger, 999.
18. Petroski, "Human Factors," 309. See also Petroski, *Invention by Design: How Engineers Get from Thought to Thing* (Cambridge, MA: Harvard University Press, 1996).
19. Ibid.
20. Thomas Hornstein, "Telephone Interfaces on the Cheap," *Proceedings of the UBILAB Conference '94*, Zurich, 1994: 146.
21. Thomas Frank, *The Conquest of Cool: Business Culture, Counterculture and the Rise of Hip Consumerism* (University of Chicago Press, 1997).

22. Jennifer Daryl Slack and J. MacGregor Wise, *Technology and Culture: A Primer*, 2nd edition, (New York: Peter Lang, 2015): 129, emphasis in original.
23. Ibid., 131. Reterritorialization "describes the process by new articulations are forged" within (and between) technological assemblages.
24. *U.S. v. Am. Tel. and Tel. Co.*, 552 F. Supp. 131, 229 (D.D.C. 1982).
25. Larry Kramer, "Activists Allege Phone Overcharges," *Washington Post*, December 12, 1978, D8.
26. Deborah Rankin, "Personal Finance: How to Cut the Telephone Bill," *New York Times*, December 14, 1980, A13.
27. Paul L. Gioia, "New York Bell to Let Customers Buy Their Phones," *New York Times*, August 12, 1982, B15.
28. Associated Press, "AT&T Lists Prices for Phones," March 30, 1983.
29. Carl Oppendahl, "Needlessly Paying Extra for Touch-Tone Dialing?" *Wall Street Journal*, September 16, 1986, A1.
30. "Phones: Buy or Rent," *Boston Globe*, May 1, 1983; Michael deCourcy Hinds, "Consumer Saturday: Savings in Owning a Phone," *New York Times*, July 4, 1981, A13; Karen Arenson "The End of One-Stop Shopping for Telephones and Services: Making Some Sense of the Divestiture," *New York Times*, December 29, 1983, D17; Christine Winter, "New Bills Are Bad News to Phone Renters," *Chicago Tribune*, April 26, 1986, 1; Carl Oppendahl, "Needlessly Paying Extra for Touch-Tone Dialing," *Wall Street Journal*, September 16, 1986, 1; Peter Kerr, "For Phone Buyers, Some New Options, *New York Times*, August 26, 1982, C3.
31. "Pacific Bell Must Show Specific Items on Bill," *San Francisco Chronicle*, February 12, 1987, 3; "Bell Atlantic Gives Rebate, Rate Cut in Massachusetts," *Washington Telecom Newswire*, May 21, 1999; "AT&T Decides to Raise the Rent," *San Francisco Chronicle*, March 5, 1986, p. 25; Keith Benman, "Settlement May Pay Indiana Phone Renters," *The Times* (Munster, IN), December 31, 2002; *Canada NewsWire*, "Bell Responds to Customer Concerns by Withdrawing Touch-Tone Proposal," July 24, 2001.
32. A.P., "Widow Rented Rotary Phone for 42 Years," December 2, 2006.
33. Patricia Aufderheide, *Communications Policy and the Public Interest: The Telecommunications Act of 1996* (Guilford Press, 1999), 1.
34. Ibid., 65.
35. Zielinski quoted in *Communications Daily*, November 18, 1994, p. 1.
    *Communications Daily* is the flagship publication of Warren Communication News, Inc., founded in Washington, D.C. in 1945. "Originally established to cover the emerging new medium of television," according to its homepage, "WCN has constantly expanded its expertise to embrace new communications fields. Today the company is acknowledged as the leading publisher of hard news, analysis and research in the fields of telecom, broadcasting, the Internet, satellites, consumer electronics and related industries." http://www.warren-news.com/aboutwarren.htm.
36. Russell Baker, "Service with a Grimace," *New York Times*, July 13, 1985.
37. Schiller, 7.
38. Andrew Pollack, "Two Settlements May Widen the Pressures on Competition," *New York Times*, January 9, 1982; "'Simple and Straightforward:' Judge Green Lifts Ban on AT&T Electronic Publishing," *Communications Daily*, July 31, 1989, 1.
39. Arenson "The End of One-Stop Shopping for Telephones and Services: Making Some Sense of the Divestiture;" David Wessel, "Divestiture; The AT&T Breakup;

A Brave New World of Choices; Consumers Find They Have to Shop Around."
*Boston Globe*, December 11, 1983.

40. Morton Mintz, "Phone Ruling Stands," *Washington Post*, January 17, 1978, A1.

41. Ronald Rosenberg, "Two Firms Take Aim at Ma Bell," *Boston Globe*, February 11, 1981; Jane Meredith Adams, "The Battle for Long-Distance Sales," *Boston Globe*, June 3, 1984.

42. deCourcy Hinds, "A Guide to Long Distance Telephone Services;" James W. Kilman, Jr., "MCI Takes Aim at AT&T Lines," *Boson Globe*, June 24, 1982; "Digital Switch to Introduce Device for Rotary Telephones," *Wall Street Journal*, October 18, 1984; Barbara L. Isenberg and Mary Smith, "Helpful Hardware; Gadgets for Phones," *New York Times*, February 25, 1982; "Cosmo Settles Charge of FTC over Phones," *Wall Street Journal*, September 23, 1986.

43. Lisa Belkin, "Consumers Resist Choice on Long Distance," *New York Times*, March 23, 1985.

44. Bruce Keppel, "Long-Distance Phone Firms in Race for Area Customers," *Los Angeles Times*, June 9, 1986.

45. Bernard L. Albert, letter to the editor, *New York Times*, October 17, 1982, A22. All quotations this paragraph.

46. "Notebook," *Communications Daily*, August 7, 1992, 4.

47. *Seinfeld*, "The Pool Guy," 118, directed by Andy Ackerman, written by Larry David and Jerry Seinfeld. In a 1994 episode, Jerry and friends are scandalized to discover that a woman Jerry is dating moonlights as a phone sex worker.

48. Movietickets.com press release, August 2, 2004. Accessed February 29, 2016. http://www.movietickets.com/press.asp?year=2004&pr_id=42.

49. Mark Andrejevic, *iSpy: Surveillance and Power in the Interactive Era* (UP of Kansas, 2007), 1ff.

50. U.S. v. Am. Tel. and Tel. Co., 552 F. Supp. 131, 229 (D.D.C. 1982). The Consent Decree defines "information services" as "the offering of a capability for generating, acquiring, storing, transforming, processing, retrieving, utilizing, or making available information which may be conveyed via telecommunications, except that such service does not include any use of any such capability for the management, control, or operation of a telecommunications system or the management of a telecommunications service." In other words, information services did not include the telephone lines and infrastructural systems that transmitted the information.

51. "Notebook," *Communications Daily*, November 15, 1988, 3; A.P., "Japan-U.S. Phone Deal," *New York Times*, April 4, 1984, D20; "Telephone Information Services Fetch A Premium In The UK and France," *Telecom Markets*, January 28, 1986.

52. John Tierney, "Porn, the Low-Slung Engine of Progress, *New York Times*, January 9, 1994, B2; Nancy Blodgett, "976 OR NINE SEVEN SEX?: Dial-a-porn vendors targets of angry parents," *American Bar Association Journal*, 74 ABAJ 30, April 1, 1988; "Bureau Consent Order: Dial-a-Porn Operator fined $50,000 by FCC Common Carrier Bureau," *Communication Daily*, November 8, 1988, 6.

53. Dennis Hevesi, "Phone Sex Lines Disconnected After a Ruling," *New York Times*, November 16, 1989, B6; Richard Hylton, "For 900 Numbers, the Racy Gives Way to the Respectible," *New York Times*, March 1, 1992, B1.

54. Hylton, "For 900 Numbers, the Racy Gives Way to the Respectible;" Larry Luxner, "FTC Settles Audiotext Fraud Case," *Telephony*, June 30, 1997.

55. Laurent Belsie, "Pay-Per-Call Services Ringing Up Lots of Flak," *Christian Science Monitor,* October 30, 1991, 6.
56. All Supreme Court briefs and decisions, quoted in Aufderheide, *Communications Policy and the Public Interest: The Telecommunications Act of 1996* (New York: Guilford Press, 1999), 197–8.

## Bibliography

Adams, Jane Meredith. "The Battle for Long-Distance Sales." *Boston Globe,* June 3, 1984.

Albert, Bernard, L., M.D. Letter to the Editor. *New York Times,* October 17, 1982.

Andrejevic, Mark, *iSpy: Surveillance and Power in the Interactive Era.* Lawrence, KA: University of Kansas Press, 2007.

Arenson, Karen. "The End of One-Stop Shopping for Telephones and Services: Making Some Sense of the Divestiture." *New York Times,* December 29, 1983.

Aronson, Sidney H. "Bell's Electrical Toy: What's the Use? The Sociology of Early Telephone Use." In *The Social Impact of the Telephone,* ed. Ithiel de Sola Pool. Cambridge, MA: MIT Press, 1977.

"AT&T Decides to Raise the Rent." *San Francisco Chronicle,* March 5, 1986.

"AT&T Lists Prices for Phones," Associated Press, March 30, 1983.

Aufderheide, Patricia. *Communications Policy and the Public Interest: The Telecommunications Act of 1996.* New York: Guilford Press, 1999.

Baker, Russell. "Service with a Grimace." *New York Times,* July 13, 1985.

Belkin, Lisa. "Consumers Resist Choice on Long Distance." *New York Times,* March 23, 1985.

"Bell Atlantic Gives Rebate, Rate Cut in Massachusetts." *Washington Telecom Newswire,* May 21, 1999.

"Bell Responds to Customer Concerns by Withdrawing Touch-Tone Proposal," *Canada NewsWire,* July 24, 2001.

Belsie, Laurent. "Pay-Per-Call Services Ringing Up Lots of Flak." *Christian Science Monitor,* October 30, 1991.

Benman, Keith. "Settlement May Pay Indiana Phone Renters." *The Times* (Munster, IN), December 31, 2002.

Blodgett, Nancy. "976 OR NINE SEVEN SEX?: Dial-a-porn vendors targets of angry parents." *American Bar Association Journal* 74 (1988): 30.

"Bureau Consent Order: Dial-a-Porn Operator fined $50,000 by FCC Common Carrier Bureau," *Communication Daily,* November 8, 1988.

Cole, Barry G., ed. *After the Break-up: Assessing the New Post-AT&T Divestiture Era.* New York: Columbia University Press, 1991.

Coll, Steve. *The Deal of the Century: The Break-Up of AT&T.* New York: Antheneum Press, 1986.

"Cosmo Settles Charge of FTC over Phones," *Wall Street Journal,* September 23, 1986.

Deninger, R. L. "Human Factors Engineering Studies of the Design and Use of Pushbutton Telephone Sets," *Bell Technical Journal,* 39 (1960): 995–1012.

DeRouchey, Bill. "History of the Button." SlideShare. March 28, 2010. http://www.slideshare.net/billder/history-of-the-button-at-sxsw. Accessed April 6, 2016.

"Digital Switch to Introduce Device for Rotary Telephones," *Wall Street Journal,* October 18, 1984.

Fischer, Claude. "'Touch Someone:' The Telephone Industry Discovers Sociability." *Technology and Culture* 29 (1987): 32–61.

Frank, Thomas. *The Conquest of Cool: Business Culture, Counterculture and the Rise of Hip Consumerism.* Chicago: University of Chicago Press, 1997.

Gioia, Paul L. "New York Bell to Let Customers Buy Their Phones." *New York Times,* August 12, 1982.

Henck, Fred. *The Slippery Slope: The Long Road to the Break-Up of AT&T.* Westport, CT: Greenwood Press, 1988.

Hevesi, Dennis. "Phone Sex Lines Disconnected After a Ruling." *New York Times,* November 16, 1989.

Hinds, Michael deCourcy. "Consumer Saturday: Savings in Owning a Phone." *New York Times,* July 4, 1981.

Hornstein, Thomas. "Telephone Interfaces on the Cheap." *Proceedings of the UBILAB Conference,* 1994.

Hylton, Richard. "For 900 Numbers, the Racy Gives Way to the Respectible." *New York Times,* March 1, 1992.

Isenberg, Barbara L. and Mary Smith. "Helpful Hardware; Gadgets for Phones." *New York Times,* February 25, 1982.

"Japan-U.S. Phone Deal," *Associated Press,* April 4, 1984.

Keppel, Bruce. "Long-Distance Phone Firms in Race for Area Customers." *Los Angeles Times,* June 9, 1986.

Kerr, Peter. "For Phone Buyers, Some New Options." *New York Times,* August 26, 1982.

Kilman, James W. Jr. "MCI Takes Aim at AT&T Lines." *Boson Globe,* June 24, 1982.

Kramer, Larry. "Activists Allege Phone Overcharges." *Washington Post,* December 12, 1978.

Lavietes, S. "Alphonse Chapanis dies at 85; was a founder of ergonomics," *New York Times,* October, 15, 2002.

Lutz, Mary C. and Alphonse Chapanis. "Expected locations of digits and letters on ten-button keysets." *Journal of Applied Psychology* 39 (1955): 314–17.

Luxner, Larry. "FTC Settles Audiotext Fraud Case." *Telephony,* June 30, 1997.

Mintz, Morton. "Phone Ruling Stands." *Washington Post,* January 17, 1978.

Mosco, Vincent. *The Digital Sublime.* Cambridge, MA: MIT Press, 2004.

Nakamura, Lisa. "What Steven Wants: Gestural Computing, Digital Manual Labor, and the Boom! Moment," *in media res: a media commons project.* March 11, 2008. Accessed October 22, 2011. http://mediacommons.futureofthebook.org/imr/2008/03/11/what-steven-wants-gestural-computing-digital-manual-labor-and-boom-moment.

"Notebook." *Communications Daily.* November 15, 1988.

Oppendahl, Carl. "Needlessly Paying Extra for Touch-Tone Dialing?" *Wall Street Journal,* September 16, 1986.

"Pacific Bell Must Show Specific Items on Bill," *San Francisco Chronicle,* February 12, 1987.

Petroski, Henry. *Invention by Design: How Engineers Get from Thought to Thing.* Cambridge, MA: Harvard University Press, 1996.

Petroski, Henry. "Human Factors." *American Scientist,* 88 (2000): 307–9.

"Phones: Buy or Rent," *Boston Globe,* May 1, 1983.

Pollack, Andrew. "Two Settlements May Widen the Pressures on Competition." *New York Times,* January 9, 1982.

Rankin, Deborah. "Personal Finance: How to Cut the Telephone Bill." *New York Times,* December 14, 1980.

Rosenberg, Ronald. "Two Firms Take Aim at Ma Bell." *Boston Globe,* February 11, 1981.

Schiller, Dan. *Digital Capitalism: Networking the Global Market System.* Cambridge, MA: MIT Press, 1999.

*Seinfeld.* "The Pool Guy." 118. Directed by Andy Ackerman. Written by Larry David and Jerry Seinfeld.

Shooshan III, Harry M., ed. *Disconnecting Bell: The Impact of the AT&T Divestiture.* Oxford, U.K.: Pergamon Press, 1984.

Simon, Sam. *After Divestiture: What the AT&T Settlement Means for Business and Residential Telephone Service.* White Plains, NY: Knowledge Industry Publications, 1985.

"'Simple and Straightforward:' Judge Green Lifts Ban on AT&T Electronic Publishing," *Communications Daily,* July 31, 1989.

Slack, Jennifer Daryl and J. MacGregor Wise. *Technology and Culture: A Primer,* 2nd edition. New York: Peter Lang, 2015.

Stone, Alan. *Wrong Number: The Break-Up of AT&T.* New York: Basic Books, 1989.

"Telephone Information Services Fetch A Premium In The UK and France," *Telecom Markets,* January 28, 1986.

Tierney, John. "Porn, the Low-Slung Engine of Progress." *New York Times,* January 9, 1994.

"The Touch-Tone® 'Dial,'" Bell Telephone Laboratories, March 1964.

*U.S. v. Am. Tel. and Tel. Co.,* 552 F. Supp. 131, 229 (D.D.C. 1982).

Wessel, David. "Divestiture; The AT&T Breakup; A Brave New World of Choices; Consumers Find They Have to Shop Around." *Boston Globe,* December 11, 1983.

"Widow Rented Rotary Phone for 42 Years," *Associated Press,* December 2, 2006.

Winter, Christine. "New Bills Are Bad News to Phone Renters." *Chicago Tribune,* April 26, 1986.

Wu, Timothy. *The Master Switch: The Rise and Fall of Information Empires.* New York: Knopf, 2010.

# 4 What's in a PIN?

## ATMs and Keypads Beyond the Telephone

PLEASE NOTE. In several days, your new automated banking card will arrive in the mail. If it is a red card with a silver stripe, your secret code will be the same as it is now. If it is a green card with a gray stripe, you must appear at your branch, with your card, to devise a new secret code. Codes based on birthdays are very popular. WARNING. Do not write down your code. Do not carry your code on your person. REMEMBER. You cannot access your account unless your code is entered properly. Know your code. Reveal your code to no one. Only your code allows you to enter the system.[1]
—Don DeLillo, *White Noise*

When Don DeLillo's *White Noise* was published in 1985, ATMs had begun to appear in DeLillo's hometown, New York City, and other financial centers, but they were still a novelty in college towns like the one where the novel takes place. The notice reproduced above arrives silently in the mail one day, presented as block text ending a chapter. Within DeLillo's dystopian satire, it is another faceless wave of data cascading into the life of his protagonist. Information overload hovers over *White Noise*, in the form of "the airborne toxic event" that triggers the novel's action. The notice about new bank cards, with its REMINDER that "only your code allows you to enter the system," is more setting than plot, although the hum of everyday life is a central theme in *White Noise* and arguably its primary subject. The notice is presented to readers, in part, as evidence of what it means to live in the U.S. as the twentieth century waned. The note serves as one example among many of the ease with which profit-thirsty corporations can take any form of communication, even a potentially romantic one, and appropriate it for commercial purposes. Letters have become notices, and names converted into numbers.

The notice arrives late in the novel, after many of the narrator's personal and professional facades have begun to crumble around him. Much earlier in the novel readers become privy to the narrator's attitude toward automated banking.

In the morning I walked to the bank. I went to the automated teller machine to check my balance. I inserted my card, entered my secret code, tapped out my request. The figure on the screen roughly corresponded to my independent estimate, feebly arrived at after long searches through documents, tormented arithmetic. Waves of relief and gratitude flowed over me. The system had blessed my life. I felt its support and approval. The system hardware, the mainframe sitting in a locked room in some distant city. What a pleasing interaction. I sensed that something of deep personal value, but not money, not that at all, had been authenticated and confirmed. A deranged person was escorted from the bank by two armed guards. The system was invisible, which made it all the more impressive, all the more disquieting to deal with. But we were in accord, at least for now. The networks, the circuits, the streams, the harmonies.[2]

For the narrator of *White Noise*, an encounter with an ATM provides not only cash, but also affirmation of his place in the world. He has taken this walk to the bank before, and a ritualistic quality may heighten the "support and approval" he finds there. Despite precedent, however, the narrator respects and fears "the system" enough to know that being granted access into it is no sure thing; otherwise, "waves of relief and gratitude" would not flow over him upon having his estimated balance confirmed. The system's acceptance and confirmation are experiences that trigger feelings of intense satisfaction, a "bless[ing]" that imbues him with a sense of "deep, personal value," which he (defensively) insists this value is about much more than money ("not that at all"). His "pleasing interaction" with the ATM is powerful enough that it moves the narrator to describe himself and "the system" with a collective pronoun: "we were in accord, at least for now." The lingering clause foreshadows the notice from the bank, two hundred and fifty pages later, and the anxiety delivered with it. The narrator covets "accord" with the system, which the subsequent note jeopardizes. His "personal value" is thrown into question when his "secret code" is in limbo.

*White Noise* is a satire of what more scholarly critiques describe as "late" capitalism's ability to appropriate everything and interpolate everyone into a total social system organized around commercial exchange.[3] The narrator's trip to the bank demonstrates his embrace of the system around him; he never fights it. Yet the novel's parody (and much of its humor) comes from his narration, delivered in an academic language of critique befitting the Chair of Hitler Studies at a Midwestern liberal arts college. Critiques of "late" capitalism usually assume or argue that people in the U.S. live more of our lives inhabiting modalities of consumption – as consumers – than ever before, and that the machinations of a capitalist system run amok are largely to blame. Like many memorable critics of capitalism, DeLillo's accounting is existential at its core. ATMs appear as monuments of depersonalization, soulless machines providing secure, selective entry points into "the system."

The satirical description in *White Noise* of "a pleasing interaction" with an ATM suggests how thoroughly the ideas and technologies comprising self-service had been absorbed into the consumer culture of the U.S. Early advertisement for ATMs appealed to the same advantages trumpeted in the first marketing campaigns for self-service shopping – savings, convenience, security, autonomy, freedom – and ATMs were the subject of several mass marketing campaigns, especially during their infancy, but tellers had never been lionized to the extent operators were, and banks never bothered promoting the new machines under the banner of self-service. Forty years after helping turn grocery stores into supermarkets, self-service had graduated from novel to basic, and by and large consumers in the U.S. had grown accustomed to the automation of everyday routines.

The phantom of the operator does not haunt the pages of *White Noise*, but traces of the privilege Bell imbued into operator service, during the telephone's formative era, may be glimpsed in a narrator who receives personal as well as financial affirmation from an ATM. The bestowal of status on early telephone subscribers flowed more from customer service relationships with operators than from the use of a new technology. Bank tellers have disappeared in *White Noise*, but the experience of being served has not lost its capacity to bestow self-worth on consumers, personal feelings that exceed if not transcend any transaction being completed. Automating a service may remove any and all interpersonal interaction from it, but it does not follow that the subsequent intrapersonal experience will be any less powerful. In *White Noise* machines have replaced human beings, not only as service providers, but as guardians of "the system," which the narrator finds "all the more impressive" for being invisible, for no longer requiring the posting of hired guards (such as tellers and operators) to allow and deny entry. (The armed guards escorting the deranged person from the bank are exceptions to the rule, whose presence underscore how smoothly the system runs.) ATMs became the first self-service kiosks to serve as entry points and safeguards to the invisible system described by DeLillo in *White Noise*. Lisa Gitelman has argued that such invisibility is evidence of social power, yet even noncontroversial technologies like the ATM, or the telephone keypad, undergo what Gitelman calls "identity crises" as their uses are being discovered, established, negotiated and accepted (or rejected).[4] Historicizing the ATM's origins illuminates multiple trajectories in its emergence as an everyday consumer labor technology.

No less than shopping or dialing, the ATM comprises a technological assemblage of representations and practices as well as material components. And like any assemblage, the ATM is to a large extent a novel combination of established elements. For instance, telephony was an essential experiential as well as infrastructural predecessor to the ATM. As discussed in Chapter 2, the home telephone number translated residential identity into a standardized alphanumerical sequence. The Anti Digit Dialing League resisted the elimination of alphabetical (and, to some, poetic) exchange names in favor of all-digit phone numbers, perceived as contributing to a "creeping numeralism" concomitant to the digitization of everyday life during the 1960s and 70s. Not only did PIN become another everyday numerical identification, but touch-tone keypads became the standard form of telephone interface in the wake of AT&T's divestiture, as described in the previous chapter. By the time banks began introducing ATMs to their customers, during the 1980s and 90s, most callers had upgraded from rotary phones to touch-tone models. Bank customers (who presumably also had telephones) were growing accustomed to using a keypad in order to transact as well as interact over the phone. The 0–9 keypad was adopted as ATM interface, and its familiarity contributed to the ATM's naturalization.

Many glosses of the ATM and its historical development have appeared in newspapers and magazines, and I draw on several in this chapter. The handful of academic articles devoted to ATMs all appear in business journals, not in the dozens of humanities and social sciences journals whose *lingua franca* is critique of capitalism and consumer culture. The few scholarly accounts of the ATM consider it primarily, if not exclusively, as a financial technology; seldom do they analyze ATMs within social, cultural or political-economic contexts. This chapter, conversely, highlights telephony as part of the cultural as well as commercial ground into which ATMs were planted. The first section draws on the history of telephone numbers, discussed in Chapters 2 and 3, to describe the emergence of checking account numbers. The following two sections situate ATM interface within the history of transaction technology, revisiting the touch-tone telephone and cash register respectively. The fourth section describes the development of the debit card. Alongside the keypad, the debit card is a technology of consumer labor in its own right, and one that has outgrown ATMs. Once "banking cards" like the one distributed to the narrator of *White Noise* matured into debit cards, consumers began using them in lieu of cash. Over the course of the twentieth century, payment technology expanded from cash and personal credit to checks and then plastic charge cards. Debit cards surpassed credit cards as an everyday payment technology, and debit cards and PIN developed into a virtually ubiquitous duo of payment technology in the U.S. The final section discusses ATM fees and surcharges. With the ATM surcharge, self-service technology graduated from lowering labor costs and enhancing productivity to the direct collection of revenue from the performance of consumer labor.

## Before the PIN: A Brief History of the Checking Account Number

After the store, the bank is often cited as the second location where routine aspects of everyday life in the U.S. were transformed by self-service.[5] Observers of commercial banking may boast that "the financial services sector has often only been rivaled by the military as an investor in, and early adopter of, technology," but when it came to consumer labor, commercial bankers trailed retail merchants and telephone companies by decades.[6] The first self-service banking technologies were ink and carbon paper, which during the 1960s holders of checking accounts began using to fill out deposit and withdrawal slips before handing them over to a teller. Deposit and withdrawal slips helped pave the way for ATMs, but they were not designed to be a technology of "distancing" on the path toward fully automated transactions.[7] Rather, deposit and withdrawal slips emerged as a byproduct of banks' decisions to automate their data processing. Developments in computer technology during World War II and the Cold War "spur[ed] the conversion from office tabulating equipment to data processing."[8] Banks began to automate their data processing before the first mainframe computers were marketed to take on such tasks. The first technologies of automation to be installed in banks' back and central offices were mechanical, not digital, and they relied on punch cards, sorters and conveyer belts to process, move and store customers' account information. Automation arrived in banks' back offices decades before ATMs appeared in the lobbies.

Don Wetzel, later credited as a "co-patentee" of the first ATM installed in a U.S. bank, worked for IBM during the 1950s and 1960s, and he recalls that banks were generally resistant to automating any aspects of their business, even data processing, let alone customer service. ATM innovators such as Wetzel were able to convince banks that machines could work in the front of their shops only after demonstrating they could be profitable in the back. Even behind the scenes, bankers were "very conservative in their operations," and manual labor was their conventional answer to data processing problems.[9] When faced with a numbers crunch or daunting deadline, banks "just brought in more people." From Wetzel's (and IBM's) perspective, the challenge with banks was to "get them off of the manual handling of data," and convince them instead to entrust their data processing to machines. According to Wetzel, when it came to investing in automation, banks were "harder to sell than the other companies [but] it turned out to be a matter of economics." Automation came with substantial costs up front, and while it had proven profitable by the 1950s to automate data processing in any number of industries, installing a processor in the back office still meant taking a leap of faith that, sooner or later, reduced payroll costs would exceed the amount of capital sunk into the machinery.

IBM was the industry leader in computerized data processors during the 1950s and 1960s. Wetzel calls his time at IBM "the 1401 era," in honor of the first of the 1400 line of mainframe processors, introduced by IBM in 1959 and retired, already a legend, in 1971. Monthly rental of a 1401 started at $2500 and was often much higher, in addition to prohibitive installation and maintenance costs.[10] To use a 1401, a company would also need to install, for instance, "special air-conditioning, humidity control, [and] false flooring so you could run wires underneath."[11] Despite the high costs, by the end of 1961 two thousand 1401s were installed in back offices across the country, "representing about one out of every four electronic stored-program computers installed by all manufacturers at that time. The number of installed 1401s peaked at more than 10,000 in the mid-1960s," and over the course of its lifetime IBM manufactured over 20,000 1401s.[12] IBM dominated the mainframe market with its 1400 line to such a degree that its competitors became known as "the seven dwarves."

The decision to install a mainframe in a bank involved an extra step. In order for a bank to begin using a 1401, the data would have to be rearranged into an appropriate format. Hence, in order to automate their data processing, banks not only had to pay for, install and maintain a processor; they also had to impose a new form of identification on their customers: the account number.

Like telephone numbers, checking account numbers were the result of back-office automation. AT&T rolled out the rotary dial gradually, and no step was more pivotal than the introduction of telephone numbers. Account numbers similarly helped set the stage for ATMs. Both new forms of identification preceded the devices on which consumers eventually began using them, the dial and ATM, yet neither was introduced explicitly as a means of preparing customers to begin using new technology. Telephone numbers were introduced as a by-product of advances in switchboard technology, which rendered it impossible for an operator to know any longer the names of even a fraction of the callers being connected; likewise, decades later, the numbering of bank accounts was a consequence of a bank's decision to automate its data processing.

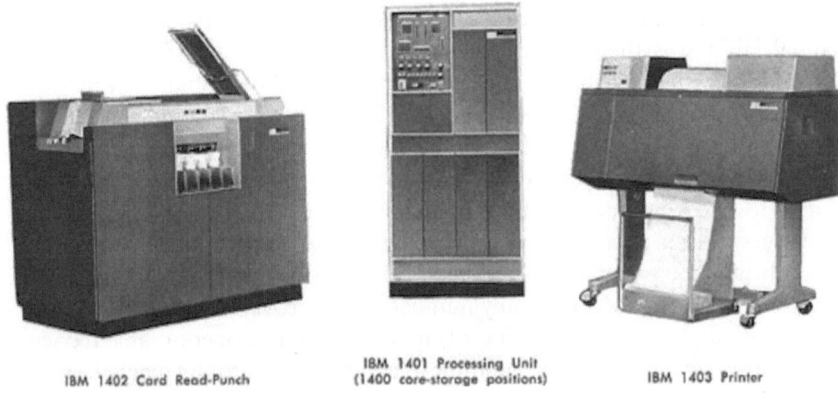

IBM 1402 Card Read-Punch

IBM 1401 Processing Unit
(1400 core-storage positions)

IBM 1403 Printer

*Figure 4.1* Data processors like IBM's 1401 needed numbers, not names.[13]

Once automation caught on, it became difficult for small or independent banks to hold out against the sea change in processing technology. During the 1960s, many bankers were still "resisting the spread of technology into their businesses. In keeping with the American tradition of small, local banks, a banker typically thought of himself in much the same way as a friendly corner grocer."[14] The parallel is ironic, given that grocers, friendly or not, had already transitioned to self-service when bankers down the street began debating whether to automate their back-office work. And even bankers who thought automation would be profitable (or was inevitable) expressed concern and skepticism about an automated future for their customer service. Unlike AT&T, who for four decades had resisted introducing *any* work into their subscribers' experience of telephony, bankers' concerns about automation were more focused. In addition to the initial costs of the machinery, bankers resisted automation, even in their back offices, primarily because of the imposition of account numbers on their customers.

Checking account numbers generated more resistance and disapproval than ATMs ever would. Alongside the retirement of exchange names in favor of "all-digit dialing," the introduction of checking account numbers during the

1960s was oft cited as an example of the "creeping numeralism" critiqued and resisted by the ADDL and its supporters.[15] Subsequently, ATM manufacturers endured more resistance from *their* customers, namely banks, than banks encountered from their own customers. (ATM surcharges have been subject of protests and lawsuits, as discussed in a later section, but never the machines themselves.) Ironically, the burden of remembering one's checking account and routing numbers, often twenty or more digits combined, would be eased by the emergence of ATM cards and their three- or four-digit personal identification numbers, or PIN. The "secret code" in *White Noise* may have figured as an ominous example of depersonalization and centralized control, but compared to lengthy account and routing numbers, PINs could feel like a relief, a reprieve of sorts from the "creeping numeralism" criticized by the ADDL in the 1960s and satirized by DeLillo twenty years later.

## Money at a Touch

Once account numbers were introduced, computer manufacturers began approaching banks about taking the next step, automating their routine customer services. (Wetzel claims the idea of an ATM occurred to him while standing in line to see a teller.) Sunk costs were a familiar deterrent, and most early ATMs came with annual costs two to three times higher than a teller's average salary.[16] Banks were also reluctant to surrender the mantle of personal service, which they allowed themselves on the grounds that any *transaction* with the bank required an *interaction* with a teller or another employee. Wetzel describes a refrain among banks at the time: "we [still] want to deal face-to-face with people. People are not going to walk up to a machine and use it. In fact we don't want them to do that, we want them coming to the bank and talking to us, because then we can sell them on other things." Although Wetzel's claim that "tellers never cross-sold to anybody" is an overstatement, ATMs expanded banks' capacity to sell their customers "on other things," including online banking.

From 2000 to 2005, online banking was the fastest growing activity among internet users, and now well over half of the people in the U.S. with a checking account conduct the majority of their banking online.[17] Banking analysts are convinced that online banking is more lucrative than ATMs.[18] On the one hand, people "visit" their bank more often online, which creates more marketing opportunities. Furthermore, people tend to do more with their money online – move it around, invest it, spend it – than they do at ATMs, and these transactions generate revenue for banks. Banks used ATMs as a bridge to the internet for their customers. Since 2005, banks have spend less money on the purchase of new ATMs than they have on upgrading older machines, networking them so they can offer more complicated services than simply withdrawals, deposits and balance checks. Upgraded ATMs often feature touch screens enabling a wider range of transactions, as well as marketing opportunities, for instance, advertisements for a bank's website – press here to learn more! Wells Fargo was one of the first banks to upgrade their ATMs, and they began offering all of their account services at ATMs, including investment brokerage and mortgage lending as well. Hardly anyone took out a mortgage while standing at an ATM, but that wasn't the point. These upgraded ATMs were often where Wells Fargo's customers' first learned that they could access these services online as well. ATMs were one of the leading promotional sites where banks informed their customers about online banking, as well as one of the sites where they taught them how to conduct it.

Banks were initially worried that ATMs would reduce opportunities to promote new products and services. Docutel, a manufacturing start-up that hired Don Wetzel away from IBM in 1968, commissioned a study from the University of Dallas business school to help prove to banks that their customers would not miss their interactions with tellers, like consumers of an earlier era had missed their telephone operators.[19] As mentioned earlier, the bank is often glossed as the second site, after the store, where everyday acts of consumption in the U.S. were transformed by self-service. A history of self-service focused on everyday technology, rather than on commercial venues, highlights the home telephone as another formative case. A focus on technology also helps illuminate why some service employees were difficult for consumers to part while others were not. The telephone was the first self-service technology to be used at home, and it was more difficult for many callers to let go of their (domestic) service relationship with operators than it was for them to begin doing parts of an operator's job, namely dialing. Saying goodbye to their "hello girls" enacted a more dramatic loss for telephone subscribers than saying goodbye to bank tellers would decades later for the holders of checking accounts.

The domestication of automated telephony was one reason for the ATM's soft landing. Another was the naturalization of self-service. As self-service spread from grocery stores and supermarkets to other retail spaces, it evolved into a common-sense concept and an everyday practice. With self-service becoming familiar to employers, employees and consumers alike, the process of preparing bank customers for ATMs and fully automated services did not to be as elaborate as AT&T's slow road to automated telephony had been. The filling out of deposit and withdrawal slips and the numbering of accounts constituted the primary acts of preparation by banks and their customers for ATMs. ("A machine to do the work of a bank teller? The idea sounded great to everyone–except bankers."[20]) Self-service had proven acceptable to shoppers in stores, and the rotary dial had turned placing a telephone call into an act of self-service, in practice if not in name. When ATMs became an option, local (or old-fashioned) bankers worried that automated transactions would narrow the distance between the customer service they provided and that offered by their corporate competitors. (In a sense, smaller banks resisting automated service resembled independent telephone companies who, conversely, around the turn of the twentieth century strove to automate their service in order to distinguish themselves from the growing AT&T monopoly.)

As companies like Docutel were beginning to design and manufacture ATMs, during the 1960s and 1970s, consumers in the U.S. were becoming familiar not only with self-service but also with push-button keypads, on touch-tone telephones and also electronic calculators. Most manufacturers competing to market early calculators, like Texas Instruments, opted to arrange the numbers from bottom-to-top; whereas AT&T, having conducted vast amounts of research and testing, decided to number the new telephone keypad from top-to-bottom. Of the two consumer electronics devises featuring push-button keypads, the calculator is arguably more closely related to the maintenance of a checking account; however, once banks began designing ATMs, there was virtually no (documented) discussion or internal debate among designers about how to arrange the keypad. The telephone's keypad, not the calculator's, was the one adopted for ATMs. AT&T devoted years of testing to decide how to arrange the numbers on their patented "Touch-Tone 'Dial'," rolled out in 1963. All that research appears to have paid dividends not only for AT&T, but for the manufacturers and operators of ATMs as well, who were able to draw on the phone company's extensive research – and the equally extensive training of their customers – in order to feature a familiar form of interface on their brand-new machines. When banks rolled out ATMs and implored their customers to adopt a "secret code" in order to use them, consumers were not only already accustomed to self-service in a variety of forms; they were also increasingly comfortable entering personalized codes into digital keypads.

*Figure 4.2* Look Familiar? It did when bank customers began using ATMs.[21]

## Cash at Hand

In 1993, there were one hundred thousand ATMs in the U.S. handling over $650 million in transactions. That November, *Wired Magazine* announced in its fifth issue that "ATMs are well on their way to becoming the soda machines of the future."[22] The comparison serves as a reminder that the history of ATMs stretches at least to the first vending machines. Over the course of the 1990s the number of ATMs in the U.S. more than doubled, as hundreds of thousands were installed in bank lobbies and other commercial sites across the country. In October, 1997 a front-page headline in the *Wall Street Journal* asked its readers: "Have you noticed all of those ATMs suddenly appearing?"[23] By 1998, more financial transactions in the U.S. were being handled by ATMs than by tellers, and by the turn of the twenty-first century ATMs were found at virtually every bank branch in the U.S., as well as many other everyday venues where cash is spent, such as bars, bodegas, supermarkets, and street corners. *Wired's* prediction has come to pass. ATMs remain statistically the most common and arguably the most iconic form of self-service kiosk in the U.S., and a variety of machines have been developed in their wake to handle a widening range of transactions – from self-check out lanes in grocery stores to self-check in at airports.

The first recorded attempt to automate retail banking services occurred in 1939, when twenty patents were filed for separate components of a new "hole in the wall" cash dispenser. After filing the patents, Luther George Simjian, a Turkish inventor who settled in the U.S. after being separated from his family during World War I, convinced what is now Citicorp to give his new machine a try, and several of them were mounted into the street-facing walls of branches throughout New York City. Simjian had already helped pioneer auto-focus cameras, and several inventions attributed to him are still in use today, including the teleprompter; however, despite his other achievements in automation, and contrary to the ATM's subsequent success story, this first cash machine was a flop. None of the branches recorded much use of it, and the experiment was discontinued after an initial six-month trial. In 1939, self-service was being adopted as a new and improved way to shop for groceries, and a majority of Americans were turning rotary dials to place local telephone calls. Automation had spilled out of the factory and was leaving its mark on domestic labor as well as office work; yet, reflecting years later, Simjian complained that the only people who used his cash machine were "prostitutes and gamblers who didn't want to deal with tellers face to face."[24] Regardless of whether Simjian grounded this rationale in observation or stereotype, the failure of his machine suggests that in 1939 round-the-clock access to cash was not yet a readily marketable service. Sixty years later, such access quickly became a standard feature of virtually any personal checking account in the U.S. The article in *Wired* echoed *White Noise* in claiming that "[o]nly when the system fails do we notice it." The banking public accepted ATMs in short order and absorbed them as routine.

After Simjian failed to market his "hole in the wall" cash dispenser, it would be more than twenty years before the next predecessor to the modern ATM appeared, again in New York City, this time at branches of First National City Bank (now Citibank). Whereas Simjian's first machine had been designed to dispense cash, clients could use the new "Bankograph," as it was called, to deposit cash or checks without seeing a teller, and even to pay their ConEd utility bills. Cash was not available for withdrawal at the Bankograph, and historians of financial technology disqualify it from inclusion as an ATM, strictly defined, on the grounds of its one-way flow of money. (The same holds for Simjian's "hole in the wall" dispenser, which did not accept deposits.) The Bankograph was also mechanically primitive in comparison to a modern ATM: for instance, an internal camera photographed each bill (or coin) fed into the machine, recorded the time and date, and roughly thirty seconds later a series of photographic copies were spat out as the customer's receipt. Furthermore, the Bankograph was not secure yet from even crude methods of theft, so its operation was limited to the branches' regular banking hours.[25] Despite these flaws and limitations, the failure of the Bankograph to catch on was not attributed to a lack of demand, as had been the case with Simjian's "hole in the wall" dispenser. In 1960, the Bankograph was retired before it was given a chance to catch on, in part because a fully operational, two way ATMs were now on its way down the pipeline, being developed by firms like Docutel, and the preparation – of bankers as well as their customers – for self-service banking had begun.

IBM dominated the market in data processing technology during the early computer era, but one of the "seven dwarves" became a major ATM manufacturer. The NCR Corporation was a competitor of IBM's, who during "the 1401 era" sold its own mainframe processor, the NCR 204, to Bank of America among other clients. The arc of NCR's career as a manufacturer of money-handling machines highlights another trajectory within the ATM's history, one stretching as far back as the telephone. Founded in 1884 in Dayton, Ohio, NCR was the first company in the world to manufacture cash registers. The previous section highlighted the home telephone, specifically the touch-tone keypad, as a key technology in the ATM's prehistory, and the cash register is another significant precursor. The cash register altered "the nature of the retail transaction," and the history of customer service at the till, as well as over the phone, comprises a significant aspect of the ATM's success story.[26]

In Chapter 1, I discussed the cash register in the context of self-service grocery shopping and described the dominance of NCR during the late nineteenth century.[27] In addition to trailblazing strategies for promoting and selling cash registers, NCR regularly upgraded the original design, for instance adding paper rolls to provide customers as well as the company with a printed receipt for each transaction. NCR also designed the first electrical cash registers as early as 1906. It would be another fifty years, however, before NCR would successfully market a second machine. During the 1920s, NCR tried and failed to make splash in the emerging typewriter market, but the company would not make the same mistake when, after World War II, American businesses began automating their data processing. By the middle of the twentieth century, NCR had expanded beyond Ohio and maintained two primary manufacturing plants, in Waterloo, Ontario, and Dundee, Scotland. As a result of its dominance of the cash register market the company had long been a blue-chipper, but "the fortunes of NCR really took off" when it entered the business of making ATMs.[28] NCR never dominated the ATM market like it did with cash registers, but the company has been a leader among ATM manufacturers from the start. Of the first million ATMs installed worldwide, over a third were made by NCR. At the turn of the twenty-first century, NCR had made one-quarter of the ATMs in use, more than any other company. By 2002, each plant was producing over 50,000 ATMs annually – the Waterloo plant for North America and Dundee for the rest of the world. In 1995, NCR opened a smaller plant in China, gaining a foothold in the growing Asian markets. At the turn of the twenty-first century NCR ATMs were dispensing cash in over a hundred different currencies in over one hundred thirty countries, totaling over $2 million dispensed every minute of every day. And, indicative of the shift from cash registers to ATMs as NCR's primary product, the original name, National Cash Register, was retired in 1974. After decades of being known to employees and clients, as well as residents of Dayton and Dundee, as "the Cash," the company is now known officially as the NCR Corporation.[29]

During the 1960s, NCR's Dundee plant developed two of the last prominent predecessors to the modern ATM. The second automated cash dispenser, after Simjian's "hole in the wall" cash dispenser failed to catch on in 1939, was introduced in England in June 1967, at a Barclays branch in Enfield, near London. "Barclaycash" vouchers were issued, free of charge, to preapproved customers who were also entrusted with a personal code number. Customers placed the vouchers in a drawer; then, the drawer closed and in response to green light (rather than a cash register's ring) the customer entered her code so the machine could check their account balance. Finally, another drawer would open containing the requested amount of cash. The following year, another "robot cashier," as they were called in England, was unveiled in London, and this one was the first to issue plastic cards for customers to use in order to activate the machine. While plastic, these "cash cards" were still essentially vouchers, which customers purchased from tellers for a predetermined amount. The cards themselves did not yet come with their own identification number, and losing one was paramount to losing cash. Furthermore, each time a customer used their card to obtain some cash, the machine kept the card for processing, and the customer could not use the machine again until the card was mailed back to them, usually several days later.

Cash registers had been around for sixty years when they began being utilized as components in self-service systems of grocery shopping, and the popularity of self-service triggered the redesign of many models, so they could be operated by deskilled labor with minimal training (and minimal chance of theft). The cash register remains by far the most common "point-of-sale terminal," as the genre is technically known, no less than the ATM remains the iconic self-service kiosk. But cash registers were designed for employees rather than customers. Despite repeated modifications intended to simplify their use, the cash register did not provide a viable model for the push-button interface of an ATM. Observers and insiders alike overlook the telephone as part of the ATM's prehistory, despite the shared interface. For instance, Don Wetzel describes ATMs as the first instance of consumers "interface[ing] directly with a piece of hardware, especially one that dealt with their money."[30] ATMs were the first everyday machine to dispense and accept cash, but subscribers used the telephone from its inception to "deal with their money," and touch-tone keypads allowed transactions to be completed over the phone.

Familiarity with self-service as well as touch-tone keypads contributed to the naturalization of ATMs as well as their proliferation. In the ATM's wake, "companies have been eager to tap into the free labor pool of customers who can be convinced to help themselves with similar machines."[31] Given that self-service emerged in grocery stores, it is no coincidence that supermarkets were one of the first venues beyond banks where ATMs were installed. The success of ATMs, in turn, spurred a proliferation of self-payment kiosks used to facilitate the direct, fully automated exchange of money for goods. Over $2 two trillion worth of retail transactions take place at self-payment kiosks annually.[32] Self-check out aisles in supermarkets are a contemporary descendant of both ATMs and cash registers. The telephone keypad migrated from the ATM to the debit card reader, and shoppers use scales and scanners virtually identical to those used by cashiers. In self-check out lanes, consumers' familiarity with ATMs as well as cash registers comes in handy.

## The Magnetism of Plastic Payment

The telephone keypad became a key component of the ATM assemblage, namely its interface. Conversely, the Personal Identification Number (PIN) and debit card (originally called an ATM card or banking card) emerged within the ATM assemblage, before maturing into significant technologies of consumer labor in their own right. When the first banking cards were issued, credit cards had been gaining popularity for twenty years. Diners Club cards began issuing cards in 1949, then during the 1950s, banks began issuing "charge cards" to some of their preferred customers.[33] "In 1966, a group of seventeen banks formed a "federation of reciprocal credit card acceptance," which eventually grew into the MasterCard networks of credit cards.[34] MasterCard, along with its companion/competitor Visa, dominated the credit card market during its formative era. It was during this era that Docutel, another of data processing's "seven dwarves," installed the first modern ATM at a bank in the U.S., a Chemical Bank branch in suburban

New York. Docutel's initial machines were "relatively simple ... put together from existing technologies that they modified for their purposes."[35] I have highlighted the touch-tone keypad has one of these existing technologies. Other companies, such as NCR in their Scotland plant, were also assembling similar elements into a cohesive machine, but the "revolutionary step" taken by Wetzel and Docutel was to put magnetic stripes on the back of plastic cards to serve as the "activating device" for their ATMs.[36] When Docutel began experimenting with "mag stripe" encoding, as the technology was known for short, some metropolitan transit authorities began using similar stripes on prepaid access cards (replacing tokens) during the 1960s. The stripes on transit swipe cards keep track of their fluctuating value, as they are used and replenished, but they have never stored personalized information like one's "secret code" or the available balance in a checking account.

ATMs were a watershed technology for cash transactions, while ATM cards helped naturalize the use of plastic to organize payment. ATMs expanded the use of plastic cards from credit purchases to "cash" transactions in two stages. First, consumers began using ATM cards to withdrawal cash from their checking accounts, which they could then use to makes purchases. Then, with the advent of debit cards, consumers began using plastic to make payments directly out of their checking accounts. The magnetic stripes on the back of debit cards are the "activating device" that allows them to network with ATMs and millions of other machines through which consumers also swipe their credit cards. Today there are "literally millions of credit- and debit-card authorization devices" around the world, and charge cards have also become "the dominant form of payment on the internet."[37]

Unlike NCR, which remains a leader in ATM manufacturing, Docutel has been called a "one-hit wonder – the Knack of the banking industry."[38] The company may have rolled out the first ATM in the U.S., but the Doucteller, as it was called, was their "My Sharona." Docutel's pivotal role in the history of self-service technology does not involve their ATM machine itself; rather, Docutel's "great contribution," not only to the history of banking, but to self-service and the automation of commerce more broadly, "was the use of a magnetically encoded plastic card – of the type that is now universal – instead of one punched with holes" or a voucher system, as in the English forebearers to the ATM mentioned earlier.[39] Docutel's magnetic stripes on the back of their ATM cards would, in turn, be placed on debit cards, which have become an even more transformative consumer technology than the ATM itself. Ironically, debit cards became a competitor of sorts for ATMs, the machines' first "rival" spending technology. Debit cards have become an increasingly accepted substitute for cash, and an increasing number of retail transactions are completed today without customers having to withdraw cash first; instead, the holders of a checking account may now use their debit cards to make payments directly upon purchase. In 2007, debit cards overtook credit cards in terms of purchase volume, and since then Visa and Mastercard have reported debit card volume increases in the U.S. of about ten percent annually.[40] An industry study called debit card use in the U.S. an "unstoppable train" contributing to the decline in cash transactions, which has been fell at a rate of fifteen percent per year throughout the first decade of the twenty-first century.[41]

Don Wetzel remembers developing the ATM itself as relatively easy compared to "the problem of what was going to be the activating device to use this machine?"[42] Not only was a device needed to activate the machine, but that device would also need to verify the user's account information, namely their balance, each time they used an ATM. Docutel needed an "activating device" that would not only enable the ATM card to store data about the cardholder's account, but also allow that data to be updated after each use. In addition to Wetzel, credited as an ATM visionary, Docutel also employed at least one engineer who had worked for the federal government and had learned "quite a bit about encryption." Depsite the informational complexities, designing and encrypting the magnetic stripes was less of a problem for Docutel than finding a manufacturer to put them on plastic cards. The issue, as Wetzel recalls, was one of scale. "If you just hand-made some plastic cards, the machine would work, but if you went to a manufacturer and asked him to run off 5,000 cards for you, most of them didn't work." "Mag stripe" technology had already been applied to some credit cards and transit cards, but neither of the credit cards giants had made a systematic conversion yet, and Docutel "had a heck of a problem finding someone who could produce, especially in volume, a card that would work." Coming up with a manufacturer for their ATM cards "almost became a calamity" for Docutel, until the firm designed its own magnetic stripe technology.

Despite pioneering the technology behind the debit card, Docutel faltered as the ATM markets matured and did not survive the 1980s. Docutel adopted a proprietary approach to its magnetic stripes landed it in direct competition with Citibank, the major player in the emerging market for ATMs in the U.S. During the early 1970s, shortly after Chemical Bank unveiled its first Docuteller and issued new plastic cards to activate it, Citibank introduced its own "Citicards." Initially Citicards were used strictly as a form of identity verification, an easier way to "access the system" than remembering or writing down one's account number, but from the start they were envisioned as part of a systematic effort on behalf of Citi's managers and executives to prepare their customers for ATMs. Unlike Chemical and other banks that purchased their ATMs from manufacturers like Docutel, Citibank was large enough and wealthy enough to build its own machines.[43] Paul Glaser, Citibank's Chief Technology Officer, cited "poor human factors" on the commercially available machines as a reason for building their own ATMs in house. Furthermore, the early machines for sale in a burgeoning ATM market "couldn't be hooked up to the central network that was already in place at Citibank, and software couldn't be installed from a central location," and this technological incompatibility also factored into Citibank's decision to develop their own ATMs.[44] When Citibank began installing its own ATMs, gradually at first, they had already issued Citicards to selected customers. At first these cards were used "only to verify personal checks," but from the start Citi's executives envisioned them for "general use by customers," for instance "to determine their balances and other account information," in addition to withdrawing cash. As one Citibank executive (obnoxiously) put it at the time, the use of Citicards would be expanded so that the bank could develop "ATMs for the great unwashed." Unlike the magnetic stripes developed by Docutel, Citibank embedded their ATM cards with a microchip that its machines (and only its machines) could read. During the 1970s, "the war between the mag stripe and the magic middle," as Citibank's technology was known, shaped the future of electronic payment in the U.S.[45]

"Citibank couldn't take credit for being the first to install cash machines, but it would be the first to do so on a massive scale," when in 1977 it "announced that it was launching the largest ATM network in the nation."[46] In a biography of Walter Wriston, Citibank's C.E.O during this era, Philip Zweig writes of the Citicard that it "was the foundation for the ATMs. ... the infrastructure on which the ATMs network would later be built and the 'training wheels' that got Citibank customers used to them."[47] Undoing the branch as the locus of consumer banking services had been a goal of Citibank executives for years, and indeed their desire to outgrow the branch system helped convince them that ATMs were the future of consumer banking, despite projections of years of debt until the machines paid off, as well as internal predictions that automating routine customer services would trigger "five to ten years of turmoil in consumer banking." These projections, and the Citi brass's willingness to absorb the costs, recall the calculations and deliberations among AT&T executives, decades earlier, when deciding whether and then how, and at what pace, to automate their telephone service. The bankers followed a similar plan to that described by Venus Green as AT&T's discursive as well as protocological "distancing" of their customers from established services. Already in the 1970s, the more progressive and outspoken of Citibank's executives were referring to brick-and-mortar branches as "corporate dinosaurs," and even the more sober or conservative among them publicly as well as privately "doubted that the future of consumer banking lay with the branches." Citi was the first bank to begin installing ATMs at locations other than its branches, which meant that Citibank's customers no longer needed to travel to a bank in order to conduct their banking.

Like the dial's gradual spread across the U.S., the popularity of Citibank's ATMs and Citicards increased only modestly at first – until, that is, they received some help from the weather.[48] Less than a year after Citibank had committed hundreds of millions of dollars to implementing the first large-scale network of ATMs, a blizzard in New York City became the first occasion when consumers used ATMs in any volume. "In 1977, Citibank announced it would blanket New York with ATMs. ... In January 1978, [the c]ity was blanketed with something else: 17 inches of snow."[49] The bank "would have been hard-pressed to plan a better media campaign to introduce New Yorkers, long accustomed to the inconveniences of nine-to-three banking, to the advantages of cash machines." Citibank's marketing division took advantage of the snowfall to produce a commercial – in less than three days – that featured "New Yorkers of every description trudging through the slush and sleet, throughout the day and night, into Citibank ATM installations."[50] When Chemical Bank installed the first ATM in the U.S., a Docuteller at a suburban Rockville Centre, New York, branch in 1969, they ran advertisements in local papers announcing that, "[o]n Sept. 2nd, our bank will open at 9:00 and never close again!" A decade later, Citibank was beginning to install ATMs not only at branches, but everywhere they could. During the blizzard the new corporate leader in automated consumer banking introduced a slogan it still uses today: "Citi Never Sleeps." The week after the blizzard, Citibank claimed that use of their ATMs increased by twenty percent. Regardless of the actual numbers, when Citibank took advantage of the blizzard of '78 to promote the advantages of self-service banking at ATMs, it left the notion of the branch behind.

Citibank first planted their flag in the new territory of consumer banking – ATMs beyond the branch – but ultimately Docutel's magnetic stripe, rather than Citibank's "magic middle," was adopted as the technology for storing and updating account information on debit cards. The magnetic stripes that Docutel pioneered were comprised of separate "tracks" on which data could be encoded. When they introduced the stripes, Docutel initiated a "proprietary scheme" for coding them, whereby they occupied all four tracks of each stripe with their own information. Other early manufacturers of ATMs (such as Diebold, which later gained notoriety in the U.S. as a manufacturer of electronic voting machines) used only "track one" on the stripes, in part, ironically, because it was alphanumeric, and therefore could store names of account holders as well as numbers. Confining one's data to one track was also an act of foresight, anticipating that other companies in other industries would eventually occupy the other tracks.

By 1981, two years after the blizzard, Citi had "more than doubled" its market share of personal checking accounts in New York City, and their "rivals had to catch up."[51] Its competitors in the city "found themselves caught off guard by Citibank's massive deployment" of ATMs, and the other banks, belatedly, began "developing shared teller networks" that would enable their customers to use one another's ATMs.[52] "Initially, each bank operated its own ATMs [but a]s banks began to install their own off-site ATMs, it became clear that sharing machines could be much more cost-effective than having every bank deploy its own ATMs, sometimes ridiculously close to one another."[53] In 1985, seven prominent banks in the city "founded a network called New York Cash Exchange (NYCE) to link the eight hundred ATMs the banks had among them." This combined number was more ATMs than Citibank itself had at the time, and the other banks hoped that joining forces would provide a shortcut to ATM success, without having to sink as many hundreds of millions of dollars into the technology as Citi had. The NYCE in 1985 marked the first standardization of ATM cards, as banks for the first time agreed to use the same track to code their data, so that customers from any of the banks could use all of the banks' ATMs.[54]

Docutel resisted the standardization, but other problems also contributed to the company's failure, including persistent bugs in their machines and substandard marketing techniques.[55] By the time Docutel went belly up, Don Wetzel had moved on. In the late 1970s, Wetzel and a former Docutel salesman started a company called Electronic Banking Systems, which became one of the first companies to sell ATMs to clients other than banks.[56] For example, Wetzel remembers Docutel being "the first company that put ATMs in supermarkets." Citibank had already tried, and failed, to use banking kiosks in supermarkets as a step away from the branch. First, the bank planned to install machines at which Citibank customers could use their Citicards to authorize checks for payment, rather than visiting a bank branch to cash or deposit them. These Citicard check-authorization stations would be "followed by ATMs that would merely dispense cash," as a further development toward fully functional two-way ATMs. "But the reception at the supermarkets was negative; for one thing, the owners wanted to get paid for cooperating."[57] Years later, after ATMs had taken root in banks, Wetzel and EBS were able to convince grocers to install ATMs by charging consumers a fee to use them and splitting the proceeds. "[W]e would charge so much a transaction and the supermarkets would get so much. That's how we started working with the switching companies like Cirrus," which took the NYCE idea to the next level. "During the 1980s, the rise of the first regional and then national ATM networks drove the ATM industry to new heights, and radically changed consumer banking."[58] Now, not only could consumers use ATM cards at banks other than their own, they could access their money from independently owned ATMs. In a sense, one of the candidates for the title "inventor of the ATM," Don Wetzel, is also one of the founding fathers of the ATM surcharge, a significant financial technology in its own right, one arguably as significant as the debit card or ATM itself.

## Self-Service Surcharge

The surcharges applied to using ATMs mark a historical turning point for consumer labor technology. The surcharge joined the ATM and the debit card as an everyday feature of commercial life in the U.S. on April, 1, 1996, a "watershed moment" in the history of not only ATMs, but of electronically mediated commerce more generally. April Fool's Day was the perfect date for the two primary national credit card networks, Visa Plus and MasterCard Cirrus to "drop their long-standing ban on customer surcharges." Until then, "the major shared networks ostensibly prohibited surcharging in an attempt to build consumer acceptance of foreign transactions," those handled by a machine not owned and operated by the bank with which the user had an account.[59] Cirrus and Plus had resisted efforts "by both banks and state legislatures" to overturn the ban, and prior to 1996, in fact, sixteen states passed laws to override it. With the Department of Justice considering antitrust charges, Cirrus and Plus voluntarily lifted the ban, and "the gold rush was on. Suddenly ATMs were no longer a mere convenience offered by banks," and banks were no longer the only companies who could turn a profit by owning and operating them. Above and beyond the changes in consumer banking already organized around ATMs, now owning the machines themselves "could be a real business with actual revenue." "Independent operators stampeded into the arena, throwing ATMs into every mom-and-pop store coast-to-coast. Over the next four years [1996–2000] the number of ATMs around the country nearly doubled"; by the turn of the century, more ATMs were owned by independent operators than by banks.[60] So, when newspaper articles, like the headline mentioned earlier from the *Wall Street Journal*, began asking readers if they had noticed "all those ATMs suddenly appearing," the more pointed question would have been, have you noticed all those ATM surcharges suddenly appearing?

Some people did notice, of course, and while I have found no organized resistance to ATMs themselves, even among bank tellers, ATM surcharges have been opposed by consumer groups; local, state and federal legislators; and even the Pentagon (advocating to have fees waived for military personnel on active duty). The surcharges became the lighting rod for protest, but they were not the first fees attached to ATM use. During the 1980s, when banks began networking their ATMs, they also began charging one another an "interchange fee" whenever one bank's customers used a machine owned by another bank. The banks claimed interchange fees were necessary to finance the networking of ATMs, and they had the added advantage of dissuading customers from using competitors' ATMs. Each time a bank's customer used an ATM belonging to someone other than their bank, the bank would also pay a "switch fee" to the network being used (i.e., Plus or Cirrus), on top of an annual membership fee to remain part of the network. "Originally, these fees were not passed onto consumers," lest they "reduce consumers' willingness to use ATMs."[61] Banks were afraid that too many fees too soon would "kill the goose that laid the golden egg," and it would not be until consumers were "essentially hooked" on ATMs that a surcharge fee for using one would become commonplace. During the mid-1990s, banks began to "argue that they needed to charge a second fee for the exact same operation, but with a different justification," and they began charging the customer, as well as the customer's bank, each time one of their ATM was used. This second fee is the notorious "surcharge." According the Office of Thrift Supervision (part of the U.S. Treasury Department), at the turn of the twenty-first century, the average ATM transaction cost the ATM operator $0.27, including sunk costs into the equipment, as well as the price of telecommunication technology comprising the bank's networks, and the labor costs of personnel in charge of ATM operations. At the same time, in 1999 banks collected nearly $2 billion dollars in interchange fees and over $2 billion more in surcharges, "a sizeable share of their profits, which were $61.9 billion in 1998."

Neither the productivity nor profitability of consumer labor is mitigated by the payment of a wage. The telephone was the first consumer labor technology that consumers paid to use, unlike the shopping cart, but they were already paying to use it when dialing transformed telephone connections into consumer labor. The work of dialing was imposed on telephone subscribers in order to cut payroll costs. As with telephone connections, automated banking was implemented to reduce labor costs, first when banks' data processing was turned over to computers, and then when customers began transacting with ATMs instead of tellers. Unlike the processors in the back offices, however, ATMs generate revenue as well as cut costs. "ATM networks have shifted from being merely a way to cut costs to becoming their own source of revenue, as various fees and new services ... are added."[62] The profits made possible by installing an ATM do not just flow from exchanging the costs of labor for the costs of machinery. With ATMs surcharges, consumers began paying outright to serve themselves.

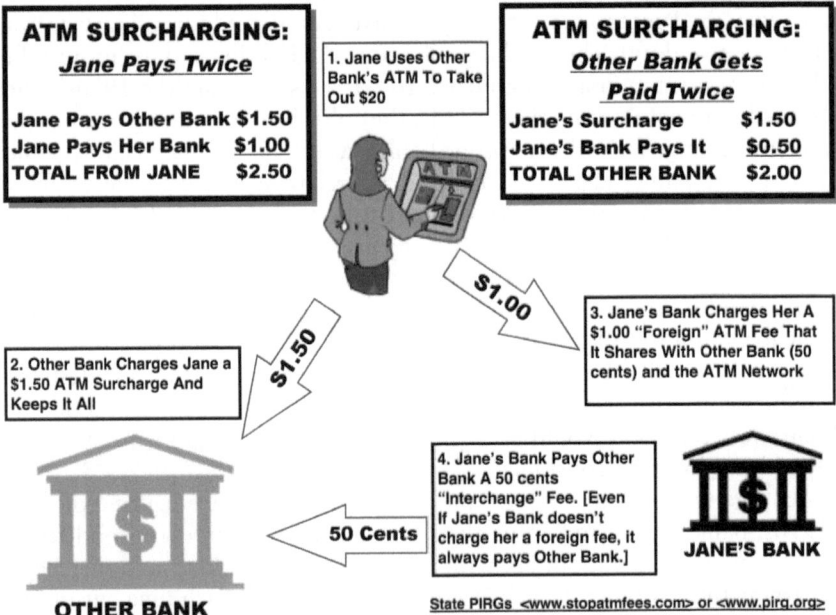

*Figure 4.3* Breaking down the costs of using an ATM.[63]

After the surcharge ban was lifted, in April 1996, the "nickel-and-dime 'em strategy" adopted by banks and ATM operators echoes the slow road to automation that AT&T had begun paving for its subscribers a century earlier.[64] Much as new tasks and responsibilities were introduced piecemeal into telephony, in anticipation of the dial, new charges and ATM fees were incrementally introduced and increased, which "irritated consumers" but on the whole did not "enrage them." The fees and surcharges were also distributed strategically; for instance, a consumer advocacy study found that, by 1998, Wells Fargo had implemented surcharges at every one of its thousands of ATMs in the state of California except one, the ATM in the state capitol building in Sacramento.

Protest and resistance organized in response to ATM surcharges were led and organized by a network of state-based Public Interest Research Groups. USPIRG is "the federation of state Public Interest Research Groups (PIRGs), [which] stands up to powerful special interests on behalf of the American public, working to win concrete results for our health and our well-being."[65] USPIRG tackles a welter of issues confronting consumers in the U.S., ranging from health care and electoral reform to media policy and affordable higher education, and I was surprised to see how prominently ATM fees figured within their broader platform for change during the late 1990s. Local consumer advocates also began opposing ATM fees shortly after the surcharge ban was lifted, in 1996, and during the final few years of the twentieth century, voters in several municipalities voted for bans or limits on ATM surcharges. For example, in 1999 California's PIRG was instrumental in passing "Amendment F," a ballot initiative banning ATM surcharges within the San Francisco city limits, which passed by more than a two-to-one margin. Voters in Santa Monica passed a similar measure, supported by reporter Robert Scheer among others, who around the same time was also becoming the mouthpiece of West LA's "Stop the Overlay" campaign, discussed at the end of Chapter 2.[66] Less than a month after the ballot initiatives passed, however, a federal judge postponed any new regulations from going into effect, until the local laws could be reconciled with the Financial Services Modernization Act, or the Gramm-Leach-Bliley Act, which also passed in November, 1999 and was subsequently ruled to override state and local legislation. Criticism of the FSMA understandably has focused on its erosion of consumer privacy and its promotion of banks' consolidation. In the wake of the FSMA, advocates like Scheer warned consumers not to "let your interest wane" in the battle with banks and the government over ATM fees, although that's essentially what happened.[67] For instance, the USPIRG website devoted to resisting and outlawing ATM surcharges, arguably the movement's epicenter, has not been updated in over fifteen years.

Banks and other financial institutions have continued to find new occasions to impose fees for the use of debit cards. A particularly draconian example occurred in Pennsylvania, where state residents began receiving unemployment benefits via debit cards, rather than being issued checks. The cards may be more convenient and flexible than checks, since they need not be deposited before the money can be spent. In fact, the state encourages recipients of unemployment benefits to use the cards because they are active immediately upon dispersal, whereas it will take at least ten days for a check to arrive in the mail. However, the state also began imposing fees for recipients who use the cards: not only the familiar $1.50 surcharge fees for payments or withdrawals, but also "forty cents for a balance inquiry. Fifty cents to have your card denied. Thirty-five cents to have your account accessed by telephone." CNN reported that "most of the 925,000 state residents who received unemployment benefits" were unaware of the fees when they opted for cards over checks.[68]

The financial collapse of September 2008 (and the election of Barak Obama) shed new light on the FSMA, and no shortage of commentators (including Obama) have cited the Act as a key precipitator of the subprime mortgage disaster and the global financial crisis triggered by it. Revisions to the Act focused on the commercial relationships that banks and other financial institutions can form with one another. However, banks' dealings with their customers also came in for new scrutiny. In September 2009, a year after the collapse, the *New York Times* published an in-depth expose titled "Overspending on Debit Cards is Painful, but Not for Banks," which began on the front page, above the fold, and continued in a full-page spread complete with charts, graphs and sidebars. "Debit," the article reports, "has essentially changed into a stealth form of credit [and] banks now make more covering overdrafts [from checking accounts] than they do on penalty fees from credit cards."[69] Two weeks to the day after running its expose about debit card fees, the *Times* ran another, more encouraging article (not on the front page, but on page one of the business section), reporting that Chase and Bank of America had announced they would "drastically overhaul their debit card programs by lowering or eliminating fees, changing the way they credit transactions and allowing customers to opt out of overdraft protection" programs, which can run as high as $40 per overdraft, even if the payment itself is for a much smaller amount.[70] Political pressure in the wake of the financial collapse has no doubt contributed to bank's willingness to revise their fee policies. At the same time, debit (and credit) cards are also facing a new "rival" spending technology, which may also help explain banks' willingness to loosen their grip on debit card overdrafts as such a powerful engine for profit.

ATMs spawned their first formidable rival in the debit card, which consumers began using to make payments in lieu of cash. Now smart phones with downloaded transaction apps are becoming all-purpose self-service technology, owned and operated by the consumers who use them to, among other applications, access checking accounts without the need of an ATM or debit card. Many independent merchants use attachment readers like SquareUp to process their own transactions, while chain stores like Wal-Mart are beginning to utilize self-check out apps that require neither a charge card nor a reader.[71] Similarly, returning to the locale where this book opened, self-check in kiosks at airports have already been pegged for obsolescence even while they are still becoming commonplace. Two thousand seven was the first year that a majority of airlines used check in kiosks, and not surprisingly, the number of customer service representatives employed by airlines has plummeted. Since 2002, the six major U.S. airlines have eliminated over a third of their combined total workforce, and the number of agents employed by airlines has been cut by more than half.[72] Self-check in kiosks dropped the cost of processing and printing a boarding pass from $3.68 when an agent did it, to $0.16 when flyers do it themselves.[73] Each airport kiosk costs tens of thousands of dollars to install, program and maintain, yet executives in the airline industry invested in the kiosks as an "interim technology."[74] The long-term goal is for every traveler to "hav[e] a kiosk in your pocket."[75] For instance, paperless boarding passes feature a two-dimensional bar code, which resembles snow on an analog TV set and is purportedly more secure than traditional one-dimensional bar codes. Travelers store the boarding passes on cell phones or other mobile devises, to be scanned before boarding the plane. The ultimate goal of the airline industry is to slash overhead as well as labor costs, by enabling travelers to board their flight without ever coming into contact with an airline-owned machine, let alone an employee. The end game for is for travelers to use their own independently owned and operated technology to perform the consumer labor of purchasing tickets and boarding flights, as well paying for parking while gone.

Inside commercial venues like stores and airports, keypad data entry on self-payment kiosks is giving way to touch screen interface. Meanwhile smart phones, personal computers, and tablets are replacing debit cards as our leading entry points into retail as well as financial networks. In the conclusion I address the emergence of smart phones as payment technology.

## Notes

1. Don DeLillo, *White Noise*, (New York: Penguin, 1986), 294–5.
2. Ibid., 46.
3. See, for example, Ernest Mandel, *Late Capitalism* (New York: Verso, 1999); and Frederic Jameson, *Postmodernism, or the Cultural Logic of Late Capitalism* (Durham, NC: Duke University Press, 1992).
4. See Lisa Gitelman, "Introduction: Media as Historical Subjects," *Always Already New: Media, History and the Data of Culture* (MIT Press, 2006); and Lisa Gitelman and Geoffrey Pingree, "What's New about New Media?," the introduction to their co-edited volume *New Media, 1740–1915* (MIT Press, 2004).
5. For example, Nicols Fox, "Volunteer Workers of the World, Unite," *New York Times*, April 9, 2005; "The most popular, accessible computer," *Toronto Star*, March 4, 2002; Ursula Huws, *The Making of a Cybertariat: Virtual Work in a Real World* (Monthly Review Press, 2003), 182. The third site of self-service in such accounts is usually the gas station, where drivers began pumping their own gas during the 1970s.
6. Dan Barnes, "Leading the IT Revolution," *The Banker Magazine*, January 2, 2006. Accessed February 29, 2016. http://www.thebanker.com/Archive/Leading-the-IT-revolution/%28language%29/eng-GB.

7. Venus Green, "Goodbye Central: Automation and the Decline of 'Personal Service' in the Bell System, 1878–1921," *Technology and Culture* 36 (1995): 912–49.

8. "NCR Corporation," *International Encyclopedia of Company Histories, Volume 30* (Famington Hills, MI: St. James Press, 2000). http://www.fundinguniverse.com/company-histories/NCR-Corporation-Company-History.html. Accessed February 29, 2016. See also Paul Edwards, *The Closed World: Computers and the Politics of Discourse in Cold War America* (Cambridge, MA: MIT Press, 1996).

9. Interview with Mr. Don Wetzel, Co-Patente [*sic*] of the Automated Teller Machine, Conducted by Dr. David K. Allison, Curator, National Museum of American History, September 21, 1995, NMAH Presidential Reception Suite. Accessed September 17, 2010. http://americanhistory.si.edu/collections/comphist/wetzel.html. All subsequent Wetzel quotes taken from this interview.

10. Press Release, IBM Data Processing Division, October 5, 1959. Accessed February 29, 2016. http://www-03.ibm.com/ibm/history/exhibits/mainframe/mainframe_PP1401.html.

11. Don Wetzel interview.

12. "IBM Mainframes," IBM Archives. Accessed February 29, 2016. http://www-03.ibm.com/ibm/history/index.html.

13. Image retrieved from IBM's online archives. Accessed February, 29, 2016. http://www-03.ibm.com/ibm/history/exhibits/mainframe/mainframe_intro.html.

14. Michael Lamm, "Dollars Ex Machina," *American Heritage's Invention and Technology Magazine*, 16 (2000). Accessed September 17, 2010. http://www.inventionandtech.com/content/dollars-ex-machina-1?page=2.

15. Another common target was the zip code. Street addresses for postal delivery date the turn of the twentieth century, but it was not until the 1960s that the U.S. postal service implemented a nationwide system of five-digit mail codes.

16. Ellen Florian, "The Money Machines," *Fortune*, July 26, 2004. Accessed February 29, 2016. http://archive.fortune.com/magazines/fortune/fortune_archive/2004/07/26/377172/index.htm.

17. *Toronto Star*, "The most popular, accessible computer"

18. Bob Tedeschi, "A.T.M.'s Pick Up Web Site Tricks," *New York Times*, March 7, 2005, B1.

19. Wetzel describes the study in interviews. I was informed by Carolyn Mauzy, Reference Librarian at the William A. Blakley Library, University of Dallas, that "a confidentiality agreement is signed between the University of Dallas and the Company so the information cannot be released to the public." Email correspondence with author.

20. This Q+A is the teaser below the headline in Michael Lamm's, "Dollars Ex Machina."

21. Photo by Christina Dunbar-Hester.

22. Phil Patton, "The Buckland Boys and Other Tales of the ATM," *Wired Magazine* 1.05 (1993). Accessed September 17, 2010. http://www.wired.com/wired/archive/1.05/atm.html.

23. Matt Murray, "Have You Noticed All of Those ATMs Suddenly Appearing?" *Wall Street Journal*, October 7, 1997, A1. For other examples of news coverage, see Dan Schiller, *Digital Capitalism*, 213 n26.

24. "Inventor of the Week" M.I.T. Program series. Accessed September 17, 2010. http://web.mit.edu/invent/iow/simjian.html. For more on Simjian, see for

example: Dan Barnes, "Leading the IT Revolution," and Michael Lamm, "Dollars Ex Machina."

25. Constance L. Hayes, "The Bottom Line on Banks," *New York Times*, April 16, 1994, Section 14, A1.

26. Kenneth Brevoort and Howard P. Marvel, "Successful Monopolization Through Predation: The National Cash Register Company," *Antitrust Law and Economics: Research in Law and Economics, Volume Twenty-One*, J.B. Kirkwood, ed., (Philadelphia: Elsevier, Inc., 2004) 89.

27. See 61–5.

28. "Fortune built on making ATMs," *The Scotsman*, January 12, 2007, 3.

29. "The most popular, accessible computer," *Toronto Star*.

30. Wetzel interview.

31. Alice Dragoon, "Six Simple Rules for Successful Self-Service," *CIO Magazine*, October 15, 2005. Accessed February 29, 2016. http://www.cio.com/article/13080/Six_Simple_Rules_For_Building_Successful_Self_Service_Application.

32. "Self-Checkout Usage Statistics," *Statistic Brain Research Institute*, last modified March 26, 2015. Accessed February 29, 2016. http://www.statisticbrain.com/store-self-checkout-usage-statistics/http://www.statisticbrain.com/store-self-checkout-usage-statistics/.

33. On Diners' Club, see Lana Swartz, "Gendered Transactions: Identity and Payment at Midcentury," *Women's Studies Quarterly* 42 (2014): 137–53.

34. David Moschella, *Consumer-Driven IT: How Users are Shaping Technology Industry Growth*, (Harvard Business School Press, 2003), 69–70.

35. "The Rise and Fall of Docutel," *ATM Marketplace*, March 18, 2003. Accessed September 17, 2010. http://www.atmmarketplace.com/article.php?id=1339&na=1.

36. Ellen Florian, "The Money Machines."

37. Moschella, *Consumer-Driven IT*, 70–2.

38. "The Rise and Fall of Docutel," *ATM Marketplace*.

39. Michael Lamm, "Dollars Ex Machina."

40. Tim Chen, "Debit Card and Credit Card Transaction Volume Statistics," *Nerd Wallet*, Accessed February 29, 2016. http://www.nerdwallet.com/blog/credit-card-data/credit-card-transaction-volume-statistics/. Debit card rates have increased much faster internationally than in the U.S.

41. "Study: U.S. consumer use of cash expected to decline by nearly $200 billion by 2015," *ATM Marketplace*, January 12, 2011. Accessed February 29, 2016. http://www.atmmarketplace.com/article_print/178763/Study-U-S-consumer-use-of-cash-expected-to-decline-by-nearly-200-billion-by-2015.

42. Wetzel interview, all quotations this paragraph.

43. This is not to suggest Citibank absorbed the development and manufacturing costs without consternation. In fact, the "prospect of losses and heavy R&D expenses for years to come," during the transition to ATMs and debit cards, led Citibank to consider eliminating the angled roof of the corporation's new building being constructed at the time in midtown Manhattan. Ultimately, the decision was made to retain "the building's most distinctive feature, one that was unique in American architecture," because the savings possible from scrapping it would amount to only $700,000, a tiny fraction of the hundreds of millions of dollars spent by Citibank on ATMs, debit cards and the reorganization of consumer banking services. See Phillip L. Zweig, *Wriston: Walter Wriston,*

*Citibank, and the Rise and Fall of American Financial Supremacy* (New York: Crown Business Publishing, 1996): 546.

44. Glaser, quoted in Ibid.
45. Ibid., 291, 294.
46. Ibid., 294.
47. Ibid.
48. Ibid., 541–7.
49. Ellen Florian, "The Money Machines."
50. Zweig, *Wriston*, 613.
51. Ellen Florian, "The Money Machines."
52. Zweig, 613–14.
53. Moschella, *Consumer-Driven IT*, 72.
54. Michael Lamm, "Dollars Ex Machina."
55. "Rise and Fall of Docutel," *ATM Marketplace.*
56. EBS was subsequently purchased by Hewlett-Packard for $13.9 billion. See Ashlee Vance, "A Difficult Combination," *New York Times*, September 23, 2009, B1.
57. Zweig, 542.
58. Moschella, *Consumer-Driven IT*, 72.
59. Christopher R. Knittel and Victor Stango, "Incompatibility, Product Attributes and Consumer Welfare: Evidence from ATMs," *BE Journal of Economic Analysis and Policy*, 8 (2008): 6. Next three quotations as well.
60. Ellen Florian, "The Money Machines."
61. "A Brief History of ATMs," State Public Interest Research Group's Campaign to End Extra ATM Fees. Accessed September 17, 2010. http://www.stopatmfees.com/newpage3.html. All quotations in this paragraph.
62. David C. Moschella, *Customer-Driven IT*, 72.
63. "A Brief History of ATMs," Reproduction in possession of author.
64. "A Brief History of ATMs," State Public Interest Research Group's Campaign to End Extra ATM Fees. The rest of the quotations in this paragraph.
65. "About Us," U.S. PIRG. Accessed February 29, 2016. http://www.uspirg.org/about-us.
66. See 110–11.
67. Robert Scheer, "Don't let your interest wane in bank battle," *Our Times Santa Monica*, November 18, 1999, 1.
68. Drew Griffin and David Fitzpatrick, CNN Special Investigations Unit, "Another Twist for the Unemployed: Debit Card Fees," CNN.com, March 13, 2009. Accessed February 29, 2016. http://www.cnn.com/2009/US/03/13/unemployment.fees/index.html.
69. Ron Lieber and Andrew Martin, "Overspending on Debit Cards Is Painful, but Not for Banks," *New York Times*, September 9, 2009, A1.
70. Ron Lieber, "Chase and Bank of America Revise Overdraft Fee Policies," *New York Times*, September 23, 2009, B1.
71. Terrence O'Brien, "Walmart testing 'Scan & Go' iPhone self-checkout app, cashiers becoming endangered species," *endgadget.com*, August 31, 2012. Accessed February 29, 2016. http://www.engadget.com/2012/08/31/walmart-testing-scan-and-go-iphone-self-checkout-app-cashiers-b/.
72. Roger Yu, "Airline service evolves intro do-it-yourself," *USA Today*, February 19, 2006, 1.

73. Alice Dragoon, "Six Simple Rules for Successful Self-Service," *CIO*, cio.com, October 15, 2005.
74. Paul Coby, SITA chairman and CIO, British Airways, quoted in "Checking In: Special Report," *Airline Business*, June 19, 2007, 1.
75. Mark Bergsrud, senior vice president for marketing programs and distribution, Continental Airlines, quoted in Susan Stellin, "Paper Is Out, Cellphones Are In," *New York Times*, March 18, 2008, C6.

## Bibliography

"About Us," U.S. PIRG. Accessed February 29, 2016. http://www.uspirg.org/about-us.

Bátiz-Lazo, Bernardo. A Brief History of the ATM: How automation changed retial banking. The Atlantic, March 25, 2015. http://www.theatlantic.com/author/bernardo-batiz-lazo/. Accessed April 6, 2016.

Bátiz-Lazo, Bernardo, Thomas Haigh, and David Stearns. How the Future Shaped the Past: The Case of the Cashless Society. *Enterprise and Society* 2014: 103–31.

Barnes, Dan. "Leading the IT Revolution." *The Banker Magazine*, January 2, 2006. Accessed February 29, 2016. http://www.thebanker.com/Archive/Leading-the-IT-revolution/%28language%29/eng-GB.

Brevoort, Kenneth and Howard P. Marvel. "Successful Monopolization Through Predation: The National Cash Register Company." *Antitrust Law and Economics: Research in Law and Economics, Volume Twenty-One*, J.B. Kirkwood, ed. Philadelphia: Elsevier, Inc., 2004.

"A Brief History of ATMs." State Public Interest Research Group's Campaign to End Extra ATM Fees. Accessed September 17, 2010. http://www.stopatmfees.com/newpage3.html.

"Checking In: Special Report." *Airline Business*, June 19, 2007.

Chen, Tim. "Debit Card and Credit Card Transaction Volume Statistics." *Nerd Wallet.com*. Accessed February 29, 2016. http://www.nerdwallet.com/blog/credit-card-data/credit-card-transaction-volume-statistics/.

DeLillo, Don. *White Noise*. New York: Penguin, 1986.

Dragoon, Alice. "Six Simple Rules for Successful Self-Service." *CIO Magazine*, October 15, 2005. Accessed February 29, 2016. http://www.cio.com/article/13080/Six_Simple_Rules_For_Building_Successful_Self_Service_Applications.

Edwards, Paul. *The Closed World: Computers and the Politics of Discourse in Cold War America*. Cambridge, MA: MIT Press, 1996.

Florian, Ellen. "The Money Machines." *Fortune*, July 26, 2004. Accessed February 29, 2016. http://archive.fortune.com/magazines/fortune/fortune_archive/2004/07/26/377172/index.htm.

"Fortune built on making ATMs." *The Scotsman*, January 12, 2007.

Fox, Nicols. "Volunteer Workers of the World, Unite." *New York Times*, April 9, 2005.

Gitelman, Lisa. *Always Already New: Media, History and the Data of Culture*. Cambridge, MA: MIT Press, 2006.

Gitelman, Lisa and Geoffrey Pingree. *New Media, 1740–1915*. Cambridge, MA: MIT Press, 2004.

Green, Venus. "Goodbye Central: Automation and the Decline of 'Personal Service' in the Bell System, 1878–1921." *Technology and Culture* 36 (1995): 912–49.

Griffin, Drew and David Fitzpatrick. "Another Twist for the Unemployed: Debit Card Fees." CNN.com, March 13, 2009. Accessed February 29, 2016. http://www.cnn.com/2009/US/03/13/unemployment.fees/index.html.

Harper, Tim and Bernardo Bátiz-Lazo. Cash Box: The Invention and Globalization of the ATM. Louisville, KY: Networld Media Group, 2013.

Hayes, Constance L. "The Bottom Line on Banks. *New York Times*, April 16, 1994.

Huws, Ursula. *The Making of a Cybertariat: Virtual Work in a Real World.* New York: Monthly Review Press, 2003.

"IBM Mainframes," IBM Archives. Accessed February 29, 2016. http://www-03.ibm.com/ibm/history/index.html.

"Interview with Mr. Don Wetzel, Co-Patente [*sic*] of the Automated Teller Machine." Conducted by Dr. David K. Allison, Curator, National Museum of American History, September 21, 1995. Accessed September 17, 2010. http://americanhistory.si.edu/collections/comphist/wetzel.html.

"Inventor of the Week." M.I.T. Program series. Accessed September 17, 2010. http://web.mit.edu/invent/iow/simjian.html.

Jameson, Frederic. *Postmodernism, or the Cultural Logic of Late Capitalism.* Durham, NC: Duke University Press, 1992.

Knittel Christopher R. and Victor Stango. "Incompatibility, Product Attributes and Consumer Welfare: Evidence from ATMs." *BE Journal of Economic Analysis and Policy* 8 (2008): 1–45.

Lamm, Michael. "Dollars Ex Machina." *American Heritage's Invention and Technology Magazine* 16 (2000). Accessed September 17, 2010. http://www.inventionandtech.com/content/dollars-ex-machina-1?page=2.

Lieber, Ron. "Chase and Bank of America Revise Overdraft Fee Policies." *New York Times*, September 23, 2009.

Lieber, Ron and Andrew Martin. "Overspending on Debit Cards Is Painful, but Not for Banks." *New York Times*, September 9, 2009.

Mandel, Ernest. *Late Capitalism.* New York: Verso, 1999.

Moschella, David. *Consumer-Driven IT: How Users are Shaping Technology Industry Growth.* Cambridge, MA: Harvard Business School Press, 2003.

"The most popular, accessible computer," *Toronto Star*, March 4, 2002.

Murray, Matt. "Have You Noticed All of Those ATMs Suddenly Appearing?" *Wall Street Journal*, October 7, 1997.

"NCR Corporation," *International Encyclopedia of Company Histories, Volume 30.* Farmington Hills, MI: St. James Press, 2000. http://www.fundinguniverse.com/company-histories/NCR-Corporation-Company-History.html. Accessed February 29, 2016.

O'Brien, Terrence. "Walmart testing 'Scan & Go' iPhone self-checkout app, cashiers becoming endangered species." *endgadget.com*, August 31, 2012. Accessed February 29, 2016. http://www.engadget.com/2012/08/31/walmart-testing-scan-and-go-iphone-self-checkout-app-cashiers-b/.

Patton, Phil. "The Buckland Boys and Other Tales of the ATM." *Wired Magazine* 1.05 (1993). Accessed September 17, 2010. http://www.wired.com/wired/archive/1.05/atm.html.

"Press Release." IBM Data Processing Division. October 5, 1959. Accessed February 29, 2016. http://www-03.ibm.com/ibm/history/exhibits/mainframe/mainframe_PP1401.html.

"The Rise and Fall of Docutel." *ATM Marketplace*, March 18, 2003. Accessed September 17, 2010. http://www.atmmarketplace.com/article.php?id=1339&na=1.

Schiller, Dan. *Digital Capitalism: Networking the Global Market System.* Cambridge, MA: MIT Press, 1999.

"Self-Checkout Usage Statistics." *Statistic Brain Research Institute.* Last modified March 26, 2015. Accessed February 29, 2016. http://www.statisticbrain.com/store-self-checkout-usage-statistics/http://www.statisticbrain.com/store-self-checkout-usage-statistics/.

Stellin, Susan. "Paper Is Out, Cellphones Are In." *New York Times*, March 18, 2008.

Vance, Ashlee. "A Difficult Combination," *New York Times*, September 23, 2009.

"Study: U.S. consumer use of cash expected to decline by nearly $200 billion by 2015." *ATM Marketplace*, January 12, 2011. Accessed February 29, 2016. http://www.atmmarketplace.com/article_print/178763/Study-U-S-consumer-use-of-cash-expected-to-decline-by-nearly-200-billion-by-2015.

Swartz, Lana. "Gendered Transactions: Identity and Payment at Midcentury." *Women's Studies Quarterly* 42 (2014): 137–53.

Tedeschi, Bob. "A.T.M.'s Pick Up Web Site Tricks." *New York Times*, March 7, 2005.

Yu, Roger. "Airline service evolves intro do-it-yourself." *USA Today*, February 19, 2006.

Zweig, Phillip L. *Wriston: Walter Wriston, Citibank, and the Rise and Fall of American Financial Supremacy.* New York: Crown Business Publishing, 1996.

# Conclusion
## Smart Phones and the Costs of Payment

The first cell phones were celebrated as technological marvels, but they were also criticized as interface nightmares. In 1995, for instance, Nicholas Negroponte warned that cell phones were in danger of being "'featured' to death." He lists several features that quickly became standard, including number storage and caller ID, while others, such as "credit card management," provided a peak at the cell phone's future. Then Negroponte turned his attention to the cell phone's most basic application. "Not only do I not want all those features; I don't want to dial the telephone at all. Why can't telephone designers understand that none of us want to dial telephones? *We want to reach people on the telephone!*" Adopting the verb from AT&T's most famous slogan may have been coincidental, but Negroponte also opts for an outmoded verb, dialing, to describe the task of placing a call. Ultimately, he concludes, "the problem of a telephone may not be in the design of the handset, but in the design of a robot secretary that can fit in your pocket."[1] Emphasizing human rather than non-human delegation, Marshall McLuhan once predicted that advances in technology would lead business executives to take on tasks previously done by their secretaries and assistants.[2] Nicols Fox, in his *New York Times* op-ed discussed in the introduction, made a broader although similarly gendered point when he observed (three decades later) that touch-tone telephones had turned consumers into "unpaid receptionists for businesses everywhere." In McLuhan's example, the issue was that technological advances would find elite professionals taking on office work previously considered beneath them. Fox's point is that reassignment of routine tasks in the social office often go unnoticed or are delegated in ways that make them difficult if not impossible to resist or refuse. When the dial was introduced on Capitol Hill, Carter Glass tried to defend himself against being "transformed into a telephone operator without compensation." Ever since, resentment toward self-service (on as well as off the job) have been animated less by the new work itself than by the conscription to replace service workers. The same technology that reorganizes paid labor *up* white- and pink-collar job ladders (McLuhan's executives performing their own secretarial functions) also reorganizes it *out* to Fox's unpaid receptionists. Cell phones have become unprecedentedly successful devises for reassignments in both directions.

No story more dramatically illustrates the significance of cell phone interface than the rise and fall of BlackBerry. More associated with business than leisure or entertainment, BlackBerry's success was largely attributed to its introduction of a full alphanumeric keyboard. No anticipatory text programs like "T9" were necessary, which made text messaging and emailing considerably faster and easier. One can imagine a BlackBerry commercial contrasting their keyboard to a competitor's keypad as textual technologies, along the same lines as the AT&T commercial discussed in the start of Chapter 3, which sarcastically touted touch-tone as "the greatest thing since the wheel." During the first five years of the twenty-first century, Blackberry conquered the cell phone market. By 2005 more than half of the cell phones in the U.S. bore the Blackberry brand. Then, in even less time than it took to take over, Blackberry crashed, eventually becoming a "source of shame" to users still stuck with one.[3]

Apple rolled out the iPhone in 2007, and in 2008 Research in Motion sold more BlackBerries than Apple did iPhones, but by the time Barack Obama took office in January 2009, the iPhone had become a "media juggernaut."[4] This phrase, quoted from the *New York Times* business section, captures the iPhone's success story on two different levels. The iPhone's launch was a media event in its own right, complete with shoppers being interviewed as they lined up overnight outside shiny Apple stores. Breathless coverage of the iPhone contributed to strong opening sales, despite mixed early reviews from users. The phrase also fits the iPhone because it became a versatile and expansive conduit for mobile (multi-) media consumption. For instance, Apple's first (award-winning) marketing campaign for the iPhone featured television commercials showcasing its ability to put the *New York Times* online archive in your pocket and at your fingertips. Apple's commercials recycled the hands-only mode of demonstration in AT&T commercial for touch-tone.[5] The two hands are again white, male, and manicured, but now they are working in harmony rather than competing. The narrator's voice (and the soft music underneath) is soothing rather than sarcastic, although at one level the message is the same: see how easy this is? AT&T's touch-tone commercial implied (wrongly) that the keypad would simply enable faster calls, while Apple wanted viewers to imagine that swiping a touch screen makes anything possible.

In the iPhone's shadow, "no credible cellphone [company] could *not* get a touch-screen phone started."[6] Even BlackBerry developed a touch-screen model, and early reviews were harsh. A *New York Times* headline asked, "No Keyboard? And You Call This a BlackBerry?" Declaring "the thumb keyboard the defining feature of a BlackBerry," the *Times* review dismissed the "concept [of] a touch screen BlackBerry" and criticized Research in Motion for its "zeal to cash in on some of that iPhone touch-screen mania."[7] In the context of the telephone keypad's history and its staying power, it is worth noting that the first touch-screen BlackBerry clicked when you press it, as an attempt to "soften the blow" of losing the keyboard by

*Figure C.1* The world (wide web) at your fingertip.

mimicking the experience of pressing actual keys.[8] The keyboard's advances over the keypad were apparent, but the keyboard's career was remarkably brief, absorbed into touch screens along with the keypad. The dial, keypad, and touch screen remain the telephone's three signature interface upgrades, attendant to the automation, digitization, and computerization of telephony. Meanwhile, the keyboard is already a footnote in the history of telephone interface, like the stylus, an afterthought already overshadowed by the touch screen.

Thanks to the commercial emergence of interactive touch screens, coupled with the popularity of apps, the smart phone is an unprecedentedly expansive and versatile consumer labor technology. The personal attachments forged with smart phones should not be overlooked or underestimated, whether they are used for business or pleasure. For whatever purposes we "reach people" (or content) on our phones, we touch them far more than often than we speak into them or hold them to our ear. "Being digital" (to borrow the title of Negroponte's bestseller) has changed how we use our fingers as well as our telephones. Take the transitive verb "swipe," which traditionally describes acts of theft and violence. To swipe something means to steal it, while taking a swipe at someone involves a punch or slap. Recently these actions have been joined if not supplanted in the popular imagination by two new meanings. Swiping a debit card through a reader has become routine across much of the wired world; and, once Apple rolled out the iPhone, swiping at smart phones quickly followed suit. The future of retail payment hinges on charge cards giving way to smart phones in order to complete transactions.

Every time a customer pays with plastic, the card issuer charges the merchant an interchange fee, now more commonly known as a swipe fee.

Swipe fees are the second highest operating expense for retail merchants, after labor costs. In the U.S., swipe fees are the highest in the industrialized world. In Europe, by comparison, swipe fees cost merchants one-eighth of what they cost merchants in the U.S. Swipe fee revenues have tripled in the U.S. in the past 10 years, while the actual cost of processing a debit or credit card transactions continues to fall. In the wake of the 2008 financial crisis, swipe fees belatedly came in for federal regulation. The Durbin amendment to the Dodd-Frank Wall Street Reform and Consumer Protection Act of 2010 limited banks to charging merchants 21 cents per charge-card transaction, plus .05% of each transaction. When swipe reform went into effect in October 2011, the average debit swipe fee on cards from covered banks dropped from 48 cents to 24 cents per transaction (the 21 cent fee plus .05% of the transaction, which is three cents on average). In 2012 these reforms saved consumers $5.8 billion and merchants $2.6 billion.[9] However, banks and credit card companies still collected over $20 billion dollars in debit card swipe fees. (Furthermore, the reforms do not cover credit card swipes, from which banks in the U.S. continue to reap over $40 billion annually.[10]) The reforms notwithstanding, in 2015 some of the largest retailers in the world, including Macy's, Target, and Office Depot filed a swipe fee lawsuit against Visa and MasterCard, the duopoly controlling over 80% of the markets in debit and credit cards. The retail giants opted out of a $7 billion settlement that would have covered over 7 million retailers nationwide (and been the largest antitrust settlement in U.S. antitrust history). A total of 15 retailers filed suit instead, claiming the settlement would have given the credit card duopoly too much freedom to raise swipe fee rates in the future.[11]

The political economy of swiping involves some of the wealthiest corporations in the world fighting over billions of dollars. The "swipe fee wars," as they have come to be called in some U.S. media accounts, continue to be waged by credit card companies and banks vs. retail merchants, with everyday consumers caught in the crossfire.[12] When companies like Wal-Mart develop a transaction app and offer it to their customers, they promote its convenience and efficiency, and the savings in labor costs are presented as being passed along to shoppers as lower prices. But the appeal of transaction apps for retailers involves their customers' satisfaction or loyalty less than proprietary claims on their financial data. Payment apps help merchants as well as banks capitalize (on) the wealth of personal information accessible via smart phones, compared to the relative paucity of data that can be mined from credit and debit cards.

Many merchants have begun to use transaction apps – like Square, which is designed to plug into iPhones and iPads specifically – in order to process their own transactions. Square does away with the swipe fee and charges merchants a higher percentage instead, 2.75% per swipe, or 3.75% for keypad entry of a credit card or checking account number. (This discrepancy is further evidence of swiping's ongoing naturalization.) Other transaction apps remove the need for card readers altogether. Peapod, an online grocer in the Northeast and Midwest, offers an app to replace product scanners as well as card readers. Shoppers can "restock household staples by scanning bar codes with their smartphones at home."[13] Meanwhile, manufacturers of point-of-sale terminals are trying to find a toehold in the mobile payment market. For example, Chase Paymentech promotes its "Future Proof terminal" as "adapt[ive] to advances in payment technology, such as so-called 'open wallets' in mobile phones."[14] The rationale behind the design, and the name-as-marketing slogan, is for Chase to assure its customers – merchants who might purchase the terminals – that they "are not going to be put in the position of playing catch-up."[15] As discussed in Chapters 1 and 4, supermarkets and airlines invested in kiosks for self-check out and self-check in, respectively, as "interim technologies," holdovers while smart phones continue evolving into "a kiosk in every pocket."[16] With the "Future Proof" terminal, Chase Paymentech is offering merchants the chance to buy a bridge from debit cards to smart phones.

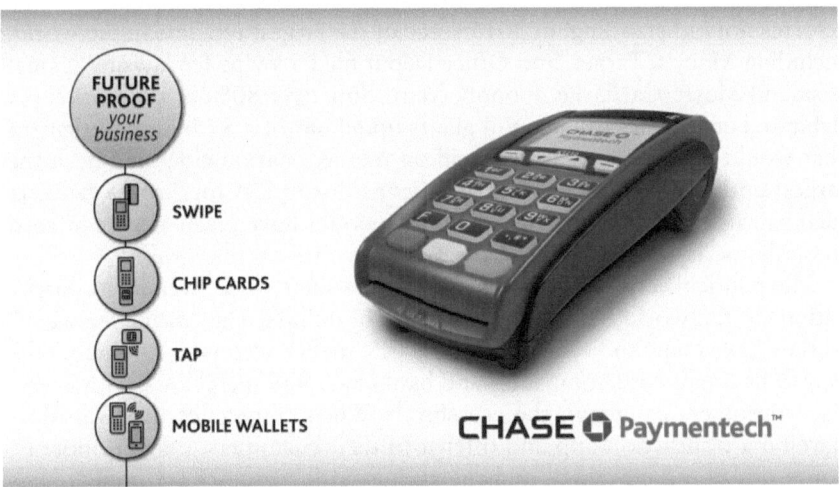

*Figure C.2* To the Future.

The telephone has been used to arrange payments and agree to prices since the days of operators and rotary dials, but the exchange of currency over the phone was not possible until monetary values could be viably, legally, and securely converted into ones and zeros. The upgrade from the rotary dial to the digital keypad was an insidious one, garnering far less attention than the dial's replacement of operators. When the dial was introduced, placing a telephone call became an act of consumer labor, but the keypad catalyzed the telephone's transformation into such an expansive and versatile consumer labor technology. The keypad also outgrew the telephone, migrating to ATMs and other digital consumer (labor) technology. Point-of-sale terminals with card swipes feature keypads virtually identical to those introduced a half-century ago on touch-tone telephones. Similarly, everyday uses for touch screens, on smart phones and beyond, continue to proliferate. Like the keypad before it, the touch screen is contributing to an expansive reorganization of telephony, in the process remediating some of the most routine tasks and transactions of everyday life.

Touch screens have been around for decades, but smart phones featuring touch screens have been widely available to the public only since 2007.[17] During Steve Jobs' presentation to shareholders that year, he unveiled the iPhone and demonstrated how to use one. Three minutes into his demo, Jobs taunted his audience by announcing "Apple has reinvented the phone, and here it is."

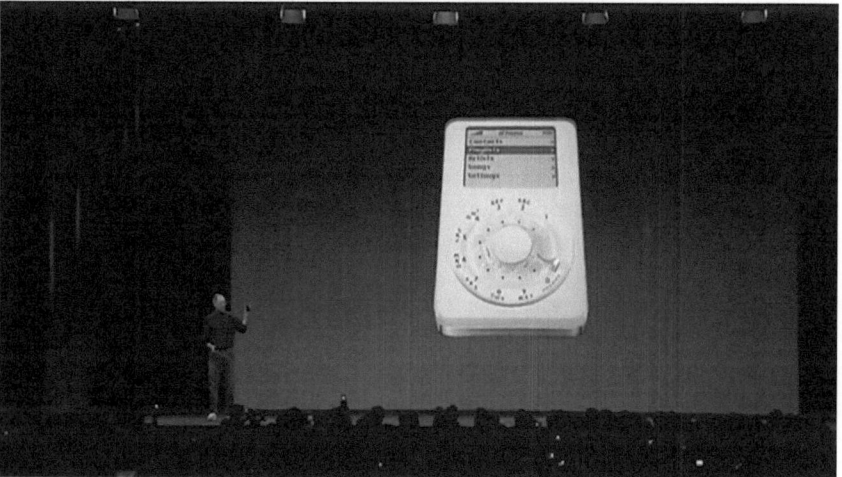

*Figure C.3* The rotary iPhone.

The very last thing Jobs demonstrated, nearly an hour later, was how to place a call. Hardly a *grand finale*, Jobs and Apple wanted placing calls to be an afterthought with the iPhone, a relic of telecommunication worthy of a condescending glance in the rearview mirror. The following year Lisa Nakamura described using an iPhone as "digital manual labor," in order to capture how quickly "the boom moment" of new consumer technology begins to wear off. Even on an iPhone, "the manual labor of interface manipulation becomes laborious soon enough, just like all the other interface required of us for work and entertainment."[18] The upgrade from the dial to the keypad is an instructive precedent for the naturalization of touch screens as interface technology. As Vincent Mosco reminds us, most transformative new media will only change the world more and more as they age. "It is when technologies such as the telephone and the computer cease to be sublime icons of mythology and enter the prosaic world of banality that become important forces of social change."[19] The rotary dial's boom moment last decades, while the telephone keypad never went boom in the first place, in the process becoming an interface technology far more powerful than the dial ever was. The joint-histories of consumer labor and telephone interface suggest that, as the touch screen's boom moment continues to fade, swiping at them will alter the everyday lives of consumers just as radically as the keypad has.

When Jobs' successor as Apple's C.E.O., Tim Cook, introduced the iWatch during his 2014 address to shareholders, he triumphantly shook his fists at the sky. But reactions to his presentation suggest that Cook's fortunes will follow, not Apple's foray into wearable technology, but rather its move into digital payment. With first the iPod and then the iPhone, Jobs turned his annual keynote address to shareholders into the media launch of the year (after Apple broke away from CES, the biannual industry-wide showcase for new tech); during Cook's first turn on the big stage, he rolled out the latest iPhone as well as unveiling the iWatch, but the biggest splash of all was made by Apple Pay. Digital systems for retail payment abound, from PayPal to Square and more recently Stripe, Braintree, and Venmo among others, but none engendered the optimism for digital payment that Apple Pay did. Cook's presentation immediately lifted the stocks of Visa and MasterCard, because he announced that Apple had not built their own closed, proprietary network from scratch, which could have cut out credit card companies entirely; rather, the company chose to capitalize on the 800 million credit and debit cards already on file with iTunes and Apple's App Store. Shares of VeriFone, the leader manufacturing of debit card readers, shot up when Cook announced that Apple would open the system to outside hardware developers – anathema to the business model single-mindedly pursued (and perhaps perfected) by Chairman Jobs. Despite some ups and downs during the initial foray into stores, within a year more than two thousand banks nationwide installed Apple Pay. Before long the U.S. government announced it would adopt Apple Pay for federal transactions like admission to national parks, for procurement cards and expense accounts issued to federal employees, and for Social Security benefits.[20]

Part of the enthusiasm for Apple's introduction of a digital payment system, as opposed to Google's digital wallet for instance, involves the relative security of their closed, proprietary networks. But rekindled hope for digital payment revolved around the sun that is Apple gadgets. Apple has long been involved with online payment, for instance via the iTunes store. The company's renowned gadgets serve as access points into their proprietary payment systems, within which Apple takes a cut from every transaction. Already, even before Apple Pay, purchase fees were Apple's second-largest revenue stream, second only to iPhone sales. "Apple actually has a really good shot at being successful here because they've solved a lot of fundamental issues that others haven't in the past. In particular, the user experience issue. It's simple, easy and secure to use."[21] The transition from charge card readers to payment apps offloads not only labor onto consumers, but also the costs of purchasing and maintaining transaction technology, now in the form of one's own gadget(s).

An even starker indication of Apple's Pay looming significance came when eBay spun off PayPal back into its own company. The cause-and-effect was captured in a *New York Times* headline: "As PayPal Spins Off, Apple Pay Signals New Era at Cash Register."[22] It is noteworthy that the cash register is invoked here as a metaphor for payment generally, rather than "the end of the cash register" being forecasted (yet again). The clunky, immovable cash register would seem to be an endangered species in a world populated by payment terminals connected to and through Apple Pay. Yet the specter of the cash register in this headline highlights the long history of interface upgrades within retail payment systems. The cash register was the first machine designed explicitly to facilitate the completion of everyday retail transactions. Cashiers use cash registers in collaboration with shoppers, instead of turning them loose to complete purchases unassisted, as we increasingly find in retail outlets today utilizing self-check out. When the routine mechanics of retail payment are reorganized, specifically around new technology from the cash register to Apple Pay, atomized tasks and responsibilities are often reassigned. For instance, debit card readers quietly shifted the task of swiping one's charge card from cashiers to shoppers.

New payment technologies organize shifts in liability as well as costs and labor among merchants and shoppers. Historically, it is the consumer, rather than the merchant, whose role expands. Recall from Chapter 1 that the first self-service lawsuits transferred much of the responsibility for shoppers' safety within a store from merchants to shoppers themselves.[23] Eighty years later, the new question about shopping is not safety, but security, specifically data security. Media coverage of recent hacks into the payment systems of store chains like Target and Home Depot (not to mention Chase bank) shone welcome light on the data collection and storage practices of retail corporations, alongside those of government agencies and telecom service providers. The credit card duopoly Visa and MasterCard set October 2015 as their deadline for merchants to upgrade to smart card technology. As described in Chapter 4, the magnetic stripes found on the back of charge cards cannot alter encoded information, rendering it much more vulnerable to capture than on chip cards, which generate unique codes for each transaction.[24] The upgrade to smart card readers costs hundreds of dollars per machine to the tune of $5 billion nationwide. Less than a year before the deadline, less than 5% of the billions of charge cards in the U.S. were equipped with chips, and even fewer readers were ready for them. Shoppers in Europe and Canada have used smart cards for years; in Europe, the deadline for chip-literate readers was New Year's Day 2005. Finally, Visa and MasterCard found enough public and political support for the upgrade in the U.S. because of heightened concern about the security of shoppers' personal and financial information. The duopoly sprang for the new cards, but they are seeing to it that costs of the new readers are borne by merchants. The cultural politics surrounding digital privacy and surveillance animate the political economy of the upgrade from striped charge cards to smart cards. After the deadline, merchants without new readers assumed liability for fraud permitted by their outmoded machines. But it's worth remembering that liability for fraud will still only fall to the merchant, rather than the card issuer, once the cardholder has been proven not guilty. As smart phones continue to replace debit cards, PIN, and readers as payment technology, what new tasks, responsibilities, and liabilities can we expect consumers to assume along the way? For starters, the costs of purchasing and maintaining one's own access nodes to payment systems. Less immediately, but more harrowing in light of the formative adjudication of self-service shopping, once consumers are accustomed to using one's own gadget to conduct retail purchases, liability for fraud may fall to consumers as well.

There have been three great leaps forward in the history of telephone interface. *Technologies of Consumer Labor* has elaborated how each – the dial, the keypad, and the touch screen – corresponds to a period of technological transformation in the telephone industry and society more broadly: automation, digitization, and computerization. Apple Pay will help usher in the fourth great leap in telephone interface, not only when biometrics like fingerprints replace PIN for identification purposes, but when the entering of one's personal data (whether by swiping a card through a reader or keying it in) gives way to that information being pulled directly from your phone. The corresponding societal trend is less financialization than datafication. Smart phones are becoming nodes of data transfer, including money, but more broadly and importantly, it seems, personal financial data like account numbers, balances, and purchase histories.

iBeacon is an "indoor proximity system" that allows ads and promos to be pinged to shoppers' phones based on their location inside (or near) a store. Futuristic marketing fantasies aside, Apple is banking on iBeacon's low-power (and low-cost) transmitters to "close the loop" of e-commerce, whereby a purchase can be publicized (and bundled for sale, etc.) the instant it is made, via the same gadget. Apple's business model involves selling consumers both content and the platforms to run it, not unlike Sony did with CDs and Walkmans, but Sony never rang up anyone's purchase or claimed a slice of the fees collected therein. With Apple Pay, Apple is becoming a *bona fide* e-commerce company, able to take a cut of not just those transactions involving their apps or iTunes, but of any transaction made with an iPhone, iPad or other Apple gadget. I, for one, was disappointed that Apple rejected iPay as the name for their digital payment system, although obviously Pay Apple would have been more accurate. The corporate control of everyday telephony has vastly broader horizons today than it did during the monopoly eras of automation and digitization.

# Notes

1. Nicholas Negroponte, *Being Digital* (New York: Knopf, 1995): 94, italics in original.
2. Marshall McLuhan, *Understanding Media: The Extensions of Man* (Cambridge, MA: MIT Press, 1994): 36. Thanks Joshua Meyrowitz for alerting me to McLuhan's prediction. On human and non-human delegation, see Bruno Latour, "Where Are the Missing Masses? The Sociology of a Few Mundane Artifacts," Wiebe E. Bijker and John Law, eds., *Shaping Technology/Building Society: Studies in Sociotechnical Change* (Cambridge, MA: MIT Press, 1992): 225–258.
3. Nicole Perlroth, "The BlackBerry as Black Sheep," *New York Times* October 16, 2012, B1.
4. Jenna Wortham and Matt Richtel, "Hoping to Draw Market Share With Touch Screens," *New York Times*, November 30, 2008, B1. President Obama was probably the most famous BlackBerry hold out. See, for example, Stephanie Clifford, "For BlackBerry, Obama's Devotion is Priceless," *New York Times* January 8, 2009, B1.
5. See 118.
6. Ed Snyder, a telecommunications industry analyst with Charter Equity Research, quoted in Wortham and Richtel, "Hoping to Draw Market Share With Touch Screens," *New York Times*, November 30, 2008, italics in original.
7. David Pogue, "No Keyboard? And You Call This a BlackBerry?," *New York Times*, November 26, 2008, B1.
8. Ibid. For other poor reviews of the BlackBerry Storm, see Rob Pegararo, "BlackBerries Again Get Sleeker but Can't Challenge iPhone," *Washington Post*, December 18, 2008, D2; Dan Frommer, "BlackBerry Phone Buyers Returning Phones En Masse?" *Silicon Alley Insider*, December 17, 2008; and Edward C. Baig, "New products that made tech fans celebrate in 2008," *USA Today*, December 17, 2008.
9. Robert Shapiro, "The Costs and Benefits of Half a Loaf: The Economic Effects of Recent Regulation of Debit Card Interchange Fees," 2. Accessed February 29, 2016. https://nrf.com/sites/default/files/The_Costs_and_Benefits_of_Half_a_Loaf.pdf.
10. Ibid., 3. Futhermore, the reforms apply to financial institutions with assets of $10 billion or more. "The 7,494 banks and credit unions with assets of less than $10 billion issued debit cards which accounted for 34 percent of 2012 debit transactions, totaling some $664 billion."
11. Anne Bucher, "Visa, Mastercard Hit with Another Swipe Fee Class Action Lawsuit," *Top Class Actions,* May 14, 2015. Accessed February 29, 2016. http://topclassactions.com/lawsuit-settlements/lawsuit-news/56002-visa-mastercard-hit-with-another-swipe-fee-class-action-lawsuit/.
12. See, for example, Maria Aspen and Gloria Finkle, "We Won, You Lost: Reactions to Credit Card Settlement," *American Banker*, July 16, 2012. Accessed February 29, 2016. http://www.americanbanker.com/bankthink/reactions-to-credit-card-swipe-fee-settlement-1050958-1.html; Karen Weise, "The Swipe Fees War Opens a New Front: Parking Meters," *Bloomberg Business*, November 5, 2012. Accessed February 29, 2012. http://www.bloomberg.com/bw/articles/2012-11-05/the-swipe-fee-wars-open-a-new-front-parking-meters; Jacob Davidson, "This is How Walmart Can Win Its War with Apple,"

*Time*, October 28, 2014. Accessed February 29, 2016. http://time.com/
money/3541247/apple-pay-walmart-current-swipe-fee-war/.
13. Hillary Stout, "For Shoppers, the Next Level of Instant Gratification," *New York
Times*, October 8, 2013, B1.
14. "Chase Paymentech Launches New Payment Terminal to Help Merchants
Securely, Conveniently Accept Newest Forms of Customer Payment," Chase
Paymentech Press Release, June 27, 2012. Accessed February 29, 2016. http://
www.reuters.com/article/idUS109308+27-Jun-2012+BW20120627.
15. Ibid.
16. See 67–70, 201.
17. Nicole Cohen, "Timeline: A History of Touch-Screen Technology," *NPR*, last
updated December 26, 2011. Accessed February 29, 2016. http://www.npr.
org/2011/12/23/144185699/timeline-a-history-of-touch-screen-technology.
18. Lisa Nakamura, "What Steven Wants: Gestural Computing, Digital Manual Labor,
and the Boom! Moment," *in media res: a media commons project*, March 11,
2008, accessed October 22, 2011. http://mediacommons.futureofthebook.org/
imr/2008/03/11/what-steven-wants-gestural-computing-digital-manual-labor-
and-boom-moment.
19. Vincent Mosco, *The Digital Sublime* (MIT Press, 2004), p. 7.
20. Andrea Peterson, "Apple Gets a Big Vote of Confidence from the U.S. Government,"
*Washington Post, The Switch blog*, February 13, 2015. Accessed February 29, 2015.
http://www.washingtonpost.com/blogs/the-switch/wp/2015/02/13/apple-pay-
gets-a-big-vote-of-confidence-from-the-u-s-government/.
21. Mike Issac, "Two Drug Chains Disable Apple Pay as a Rival Makes Plans,"
*New York Times*, October 27, 2014, B2.
22. Mike Issac, "As PayPal Spins Off, Apple Pay Signals New Era at Cash Register,"
*New York Times*, October 1, 2014, A1.
23. See 54–9.
24. See 185–9.

## Bibliography

Aspen, Maria and Gloria Finkle. "We Won, You Lost: Reactions to Credit Card
Settlement." *American Banker*, July 16, 2012. Accessed February 29, 2016.
http://www.americanbanker.com/bankthink/reactions-to-credit-card-swipe-fee-
settlement-1050958-1.html.
Bucher, Anne. "Visa, Mastercard Hit with Another Swipe Fee Class Action
Lawsuit." *Top Class Actions*, May 14, 2015. Accessed February 29, 2016.
http://topclassactions.com/lawsuit-settlements/lawsuit-news/56002-visa-
mastercard-hit-with-another-swipe-fee-class-action-lawsuit/.
"Chase Paymentech Launches New Payment Terminal to Help Merchants Securely,
Conveniently Accept Newest Forms of Customer Payment," Chase Paymentech
Press Release, June 27, 2012. Accessed February 29, 2016. http://www.reuters.
com/article/idUS109308+27-Jun-2012+BW20120627.
Clifford, Stephanie. "For BlackBerry, Obama's Devotion is Priceless," *New York
Times* January 8, 2009.
Cohen, Nicole. "Timeline: A History of Touch-Screen Technology." *NPR*. Last
updated December 26, 2011. Accessed February 29, 2016. http://www.npr.
org/2011/12/23/144185699/timeline-a-history-of-touch-screen-technology.

Davidson, Jacob. "This is How Walmart Can Win Its War with Apple," *Time*, October 28, 2014. Accessed February 29, 2016. http://time.com/money/3541247/apple-pay-walmart-current-swipe-fee-war/.

Issac, Mike. "As PayPal Spins Off, Apple Pay Signals New Era at Cash Register." *New York Times*, October 1, 2014.

Issac, Mike. "Two Drug Chains Disable Apple Pay as a Rival Makes Plans." *New York Times*, October 27, 2014.

Latour, Bruno. "Where Are the Missing Masses? The Sociology of a Few Mundane Artifacts," in *Shaping Technology/Building Society: Studies in Sociotechnical Change*, Wiebe E. Bijker and John Law, eds. Cambridge, MA: MIT Press, 1992: 225–58.

McLuhan, Marshall. *Understanding Media: The Extensions of Man*. Cambridge, MA: MIT Press, 1994.

Mosco, Vincent. *The Digital Sublime*. Cambridge, MA: MIT Press, 2004.

Nakamura, Lisa. "What Steven Wants: Gestural Computing, Digital Manual Labor, and the Boom! Moment." *in media res: a media commons project*. March 11, 2008. Accessed October 22, 2011. http://mediacommons.futureofthebook.org/imr/2008/03/11/what-steven-wants-gestural-computing-digital-manual-labor-and-boom-moment.

Negroponte, Nicholas. *Being Digital*. New York: Knopf, 1995.

Perlroth, Nicole. "The BlackBerry as Black Sheep," *New York Times* October 16, 2012.

Peterson Andrea. "Apple Gets a Big Vote of Confidence from the U.S. Government." *Washington Post, The Switch blog*. February 13, 2015. Accessed February 29, 2015. http://www.washingtonpost.com/blogs/the-switch/wp/2015/02/13/apple-pay-gets-a-big-vote-of-confidence-from-the-u-s-government/.

Pogue, David. "No Keyboard? And You Call This a BlackBerry?" *New York Times*, November 26, 2008.

Shapiro, Robert. Sonecon, LLC. "The Costs and Benefits of Half a Loaf: The Economic Effects of Recent Regulation of Debit Card Interchange Fees." October 1, 2013. Accessed February 29, 2016. https://nrf.com/sites/default/files/The_Costs_and_Benefits_of_Half_a_Loaf.pdf.

Stout, Hillary. "For Shoppers, the Next Level of Instant Gratification." *New York Times*, October 8, 2013.

Weise, Karen. "The Swipe Fees War Opens a New Front: Parking Meters." *Bloomberg Business*, November 5, 2012. Accessed February 29, 2012. http://www.bloomberg.com/bw/articles/2012-11-05/the-swipe-fee-wars-open-a-new-front-parking-meters.

Wortham Jenna and Matt Richtel. "Hoping to Draw Market Share With Touch Screens." *New York Times*, November 30, 2008.

# Index

Locators in *italics* refer to illustrations while those in bold refer to area of major coverage